Forest of Time

Forest of Time

A Century of Science at Wind River Experimental Forest

Margaret Herring

Sarah Greene

Foreword by William G. Robbins

Oregon State University Press
Corvallis

Cover photograph courtesy of the Wind River Canopy Crane
Research Facility Image Archive

The paper in this book meets the guidelines for permanence and
durability of the Committee on Production Guidelines for Book
Longevity of the Council on Library Resources and the minimum
requirements of the American National Standard for Permanence of
Paper for Printed Library Materials Z39.48-1984.

Library of Congress Cataloging-in-Publication Data
Herring, Margaret J.
  Forest of time : a century of science at Wind River Experimental
Forest / Margaret Herring and Sarah Greene ; foreword by William
G. Robbins.
      p. cm.
  Includes bibliographical references and index.
  ISBN-13: 978-0-87071-185-5 (alk. paper)
  ISBN-10: 0-87071-185-7 (alk. paper)
  1. Experimental forests--Washington (State)--History. 2. Douglas
fir--Washington (State)--Wind River Experimental Forest--History.
3. Wind River Experimental Forest (Wash.)--History. I. Greene,
Sarah. II. Title. III. Title: Century of science at Wind River
Experimental Forest.
  SD359.W55H47 2007
  634.9072'0797--dc22

                                                        2006039344

Oregon State University Press
500 Kerr Administration
Corvallis OR 97331-2122
541-737-3166 • fax 541-737-3170
http://oregonstate.edu/dept/press

Contents

Foreword

Through its long and distinguished history as the nation's principal forestry agency, the U.S. Forest Service has carried out three primary responsibilities: managing the national forest system, conducting research, and administering cooperative programs with states and the private sector. From the relatively limited research programs of the old Division of Forestry in the 1890s, federal forestry expanded its research activities during Gifford Pinchot's tenure as chief of the Bureau of Forestry when, in 1898, he created the Section on Special Investigations, a research arm elevated to division status in 1902. With the transfer of the forest reserves to the newly minted Forest Service in 1905, Pinchot, fearing that the agency was losing scientific perspective, directed staff to give more attention to research. In short order, the Forest Service's Office of Silvics, charged with directing research programs, established its first field experiment station in 1908 near Flagstaff, Arizona. In that same year, the agency sent Thornton T. Munger, a young Yale forestry graduate, to District Six in the Pacific Northwest to conduct ponderosa pine investigations. Munger, who eventually stayed in the district, represented "a one-man research unit" who shortly found "a natural laboratory in the Wind River area."

Margaret Herring and Sarah Greene's thoroughly researched *Forest of Time* traces the 100-year story of the Wind River Experimental Forest, one of several dozen such landscapes across the United States devoted to forest and range research—and the first in the Douglas fir bioregion. The authors note that for the last century, Forest Service scientists have studied Wind River's Douglas fir forests to "improve upon nature, and be humbled by it." Using a wealth of primary materials—Pacific Northwest Research Station reports, proceedings of meetings, oral interviews, agency correspondence files, annual reports, and public addresses—the authors skillfully integrate these local sources into the larger fabric of secondary literature to tell the story of the Wind River Experimental Forest. The result is a case study of a century of forest science, "of discovery and blindness, of opportunities taken and missed," with the Wind River forest in Washington's southern Cascade Range providing a spectacular natural setting for the telling of that story.

The history of forest science fits within the larger framework of the evolution of the natural sciences in the 20th century. Beginning with an exclusive focus on maximizing timber production when Munger first arrived, Wind River scientists worked to tame the wildness in the surrounding forest, grow monocrops of Douglas fir faster, and protect the trees from fire, bugs, and diseases. But nature refused to cooperate on occasion, and when scientists thought they were successful, new problems kept surfacing. With the passage of time—and through decades of change and careful observation—scientists began to develop a greater knowledge of forest ecosystems and gain a better understanding of the complex and diverse relationships linking trees, soil, water, and terrestrial life. But through this long century, there was never a clear, linear route to the present. Moreover, with every new discovery and a growing belief that researchers were making progress, there were always dissenting voices.

Herring and Greene tell this story well, weaving together the intricate strands of the sometimes contradictory research findings at Wind River and their broader implications for forest ecosystems. Although Forest Service scientists were still pursuing the elusive question of sustainability at the close of the 20th century, the meaning of the term had shifted from an exclusive focus on timber production to healthy forest ecosystems and biological diversity. Studies of sustainability began with Munger's investigations of tree growth to maximize production and continued through the end of the century, when canopy scientists began to analyze the structure and function of the forest "from roots to treetops." The work of recent researchers illustrates the very different questions and the very different ways scientists have looked at the forest from one generation to the next. The authors correctly point out that "assumptions and values changed, sometimes reversing themselves in a very short time."

The activities carried out at the Wind River Experimental Forest are a tribute to an agency whose funding has always reflected the political tenor of the times. When the focus was exclusively on production measured in board feet of timber, scientists worked hard to carry out their assigned directives—*always* with an eye to the unexpected and the unanticipated turn of events. The natural processes of growth and decay at Wind River humbled scientists who were confident they could improve on Mother Nature and easily transform the wild Douglas fir forest into a predictable agricultural crop. The findings in *Forest of Time* will surprise readers, just as the results of long-term studies carried out at Wind River surprised forest scientists.

William G. Robbins
Distinguished Professor Emeritus of History
Oregon State University

Acknowledgements

In 1908 the Forest Service was a brand-new agency and forest science was just beginning in the United States. One of the first places the agency dedicated for research was the Wind River Experimental Forest in southern Washington. Looking back on a century at Wind River, we see science changing in the same way a forest changes, both in slow growth and in cataclysmic event.

Many people helped us understand the perspective that can be gained by examining 100 years of science. We would especially like to thank Dean DeBell, Jerry Franklin, William Robbins, and Fred Swanson for their insightful and helpful reviews of the manuscript. Others who provided much-needed assistance were Cheryl Mack and Rick McClure, archaeologists, Gifford Pinchot National Forest; Cheryl Oakes, librarian, Forest History Society; Robert Curtis, mensurationist emeritus, Pacific Northwest Research Station; Theresa Valentine, spatial information manager, Pacific Northwest Research Station; Ken Bible, site director, Wind River Canopy Crane Research Facility; and Jerry Williams, retired historian, USDA Forest Service.

Introduction

✂

*Shortly after four o'clock the ash and fir needles began to
fall. A distinct roar could be heard like the machinery of
some great factory . . . Gradually the sky lightened as the fire
neared.*
　　　　　　　　　　—Eyewitness to the Yacolt Fire[1]

The day was warm and clear on September 11, 1902. No measurable
rain had fallen for two and a half months, and small fires burned in the
Cascade Mountains of Washington. Wispy plumes of smoke marked
spots where Indians dried huckleberries and settlers grubbed out farms
amid the giant forests. Fires smoldered in piles of broken limbs left
behind by loggers. The Government Land Office, which nominally
managed much of the Cascade Mountain forests, had just published a
circular describing how fire patrols should be conducted.

The wind blows hard through the Columbia Gorge, where time has
carved a narrow slot in the volcanic mountains of the Cascade Range.
Along most of the spine of the Cascades, mountains block wind and
moisture blowing in from the ocean. Peaks capture precipitation,
and rain douses the wet west slope of the Cascades, where some of
the largest forests in the world grow. But through the notch of the
Columbia River Gorge, winds blow with particular force (fig. 1).
Through this narrow gorge, moist air funnels from the Pacific Ocean
toward the dry plateau of eastern Washington and Oregon. Sometimes,
with little warning, the wind shifts. When air pressure differences are
especially strong from east to west, the wind pulls hot, dry air from
the arid plateau toward the coast and into the forested mountains.
Wind speeds can approach hurricane force, desiccating landscapes
along the way. The power of these winds can explode brush fires into
an inferno.

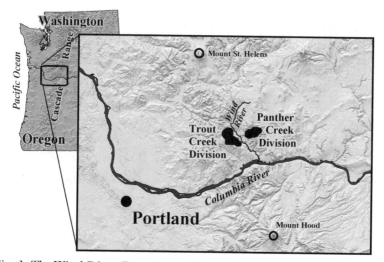

Fig. 1. The Wind River Experimental Forest, located in southwestern Washington between the Columbia River and Mount St. Helens, flanks Wind River in two divisions. (Map by Theresa Valentine, USDA Forest Service, PNW Research Station.)

On this day in 1902, sudden east winds fanned small fires throughout southern Washington. A fire in the Wind River valley merged with fires blowing in from the east, and within hours more than 80 small fires had exploded into mammoth proportions. Eyewitness accounts describe a wall of flame that thundered like a freight train, great rolling waves of flame that leapt hundreds of feet into the air, and a fierce wind that flattened young trees to the ground before incinerating them. Survivors described the deafening roar of the firestorm with its curious rhythm that beat like the heart of a monster. For three days firefighters battled the inferno, digging fire lines and backfiring day and night. Wind River valley farmers and villagers hauled water in bucket brigades to douse the monster flames. For weeks after the fire, the sound of falling trees echoed across the scorched landscape[2] (fig. 2).

Modern fire models estimate that the fire carried in the forest canopy for 36 hours at speeds up to 87 mph, with a flame length of 30 feet and a heat intensity of more than 1,000°F.[3] Over 700,000 acres burned that day in Washington alone. The devastation and loss of life was highest in southern Washington between the Wind River valley

Fig. 2. The 1902 Yacolt Burn in the vicinity of the Wind River Experiment Station, 1913. (Courtesy of USDA Forest Service.)

and the town of Yacolt, which lent its name to the conflagrations. Thirty-eight people died in the Yacolt fire, and hundreds were left homeless.[4]

Parts of the forest that had burned during the Yacolt fire ignited again over the next several decades, just as the forest had burned and reburned in patches for millennia. This is a place of wind and mountains, fire and rain. It is at the heart of a larger region—the western slope of the Cascade Range reaching from northern California to southern British Columbia, a region defined by Douglas fir. It was here at Wind River in southern Washington that the first Forest Service scientists came to learn the secrets of the Douglas fir forest, improve upon nature, and be humbled by it.

Both the Forest Service and forest science in America were brand-new at the beginning of the 20th century. When the first Forest Service researchers arrived at Wind River in 1908, the United States had recently celebrated the 100-year anniversary of Lewis and Clark's return from that same fabled, tree-shrouded region. Lewis and Clark's success helped stake the U.S. claim to the Pacific Northwest. Now the U.S. government was staking a claim for science-based management of the Pacific Northwest forests.

This book follows a century of science in a forest marked by the process of change. Focusing on the Wind River Experimental Forest (fig. 1), we follow the first Forest Service researchers who traveled across a continent to carve out new knowledge from the northwest forests. Over time, as social and technological changes transformed the 20th century, both forest science and the forest itself changed.

Those early Wind River scientists faced a wild forest of massive old trees, wastefully harvested and frequently burned, that were nothing like the cultivated woodlands of Europe and the cutover forests of eastern North America. They brought with them the revolutionary idea that forests could be tamed from that wildness and protected from fire and wasteful logging, that forests could be scientifically managed to produce sustainable crops of timber. Over time they found ways to grow timber that seemed to rival what nature herself could produce. They battled forest pests and disease and eliminated tree species that did not produce the best timber. The science they pursued at Wind River helped transform diverse wild forests into cultivated plantations of timber that were all the same age and all the same species, commercially valuable, fast-growing, and efficient to harvest.

Their successes created new problems in the forests of the Pacific Northwest. By the mid-20th century, as wild forests were giving way to plantations and as clearcuts etched deeper into the landscape, forests no longer looked like what people expected. Studies of forest ecosystems began to uncover links among forests, soil, water, wildlife, and air. Research revealed complex interactions among forest processes such as fire, disease, decay, and regrowth. Some wildlife species verged on the brink of extinction, and the rate of timber harvest, especially in old-growth forests, aroused public concern. Science became embroiled in a revolution of forest policy that led to a dramatic reduction of timber harvest on federal lands.

By the end of the 20th century, the question of forest sustainability was still at hand. Would it ever be possible to harvest timber and sustain forests in perpetuity? Could plantations be reengineered into complex forests with a mix of ages, species, and habitats? What knowledge would be necessary to manage federal forests in the future? What could 100 years of science in an experimental forest teach us about the future?

The Wind River Experimental Forest is a place, a process, a product, a laboratory, a textbook. Throughout the course of the 20th century, the science conducted at Wind River expanded and contracted, split and intertwined as generations of researchers developed new, sometimes conflicting assumptions about the forests in the Pacific Northwest. Science is a story of discovery and blindness, of opportunities taken and missed, and the Wind River forest provides a stage to tell that story. Forest scientists from a hundred years ago were just as insightful and just as blind as their counterparts today. The past was not a simpler time. Fast-forwarding through a century of science at Wind River, we are humbled by what each generation of scientists learned and accomplished within the context of their times.

Condensing a hundred years into a small book means that we have left out some events and people. Not every idea important to forest science was discovered at Wind River, but by examining this experimental forest, we can begin to understand how time changes the way we study forest environments. It is a story about how knowledge grows, how assumptions change, and how conclusions can be overturned. When the first scientists arrived at Wind River in 1908, their research focused on fighting fire and replanting burned land. By the end of the 20th century, research at Wind River focused on biological diversity and forest dynamics. New concepts have struggled to take root, buffeted by changing ideas about the environment and the shifting winds of public values. This story focuses on forest research, not agency management or politics. Forest science has gone through many iterations since 1908. There has never been a clear path to knowledge; there have always been dissenting voices. Debates continue. That is the strength of science and the underlying lesson from the Wind River story.

The Wind River Experimental Forest is one of 77 experimental forests on federal lands throughout the United States, and the oldest in the Douglas fir region. It flanks the middle reaches of Wind River in a rugged section of Washington's southern Cascades. A geologic history of volcanism shaped most of the Wind River watershed, sculpting narrow valleys with steep slopes reaching up to 4,200 feet. The boundaries of the 10,350-acre experimental forest include two

Fig. 3. A typical old-growth Douglas fir forest features tall, straight-trunked trees with layered canopies. (Courtesy of USDA Forest Service.)

divisions, one east and one west of the south-flowing Wind River, inscribing lands burned in the Yacolt fire and older forests dating back to fires in the 1600s and 1800s. The forest contains trees typical of the western slopes of the Cascade Range, including western hemlock, western red cedar, and bigleaf maple. The soils derived from the region's volcanic history are not particularly fertile, and cold air drains down from surrounding slopes, making Wind River a low-quality site for growing trees. Throughout most of the landscape, the forest is dominated by Douglas fir.

Douglas fir typically grows to 200 feet in height (some grow to 300 feet), with diameters up to 6 feet or more (fig. 3). Some live as long as 1,000 years. The trees grow straight for most of their height, eventually producing massive trunks of clear, strong wood. The great strength of the wood makes the lumber highly prized for building, and

Douglas fir overshadowed other trees in the eyes of early Northwest lumbermen. Western hemlock, which grows in the shadows of Douglas fir, was considered a weed when compared to the straight-grained wood of the Douglas fir. Western red cedar was generally dismissed by lumbermen, although it had been a basis of the regional Indian economy and vital to early settlers who found its wood easy to split and resistant to rot. Bigleaf maple, red alder, and western yew would each in time be the focus of some specialized interest, but it was the Douglas fir that would come to define the region west of the Cascade Mountains and that the Northwest timber industry would depend on for most of the 20th century.

David Douglas, the Scottish naturalist who lent his name to the Douglas fir, was not the first European to notice the majestic tree. During the 18th century, European explorers and naturalists circled the globe to collect information about faraway lands that powers back home wished to influence. Others before Douglas had ventured up the forested slopes of the Pacific Northwest, including the surgeon-naturalist Archibald Menzies, who encountered the tree in 1792 on Vancouver Island, British Columbia.[5] Meriwether Lewis made a careful study of the tree a few years later during his overland expedition with William Clark to the Pacific Northwest. During the long winter of 1806 that Lewis and Clark spent along the lower Columbia River, Lewis identified six different evergreen conifers. Describing Douglas fir, he wrote, "No. 5 is a species of fir, the bark thin, dark brown, much divided with small longitudinal interstices and sometimes scaling off in thin rolling flakes. It affords little rosin and the wood is reddish white 2/3s of the diameter in the center, the balance white, somewhat porous and tough."[6]

During an 1824 expedition with the Hudson's Bay Company, David Douglas collected samples of the tree Lewis had described and sent them to London, where the Royal Botanical Society labeled them as Douglas pine. Naturalist John Muir referred to it as Douglas spruce.[7] Loggers called it red fir or yellow fir, reflecting the two tones of the wood. But the tree was not pine, spruce, nor fir. In scientific circles, it was christened *Pseudotsuga* (false hemlock) *menziesii* in honor of Menzies, who had provided a description but had not bothered to provide a name.[8]

In any case, it was not fir but *fur* that first lured Europeans to the northwest edge of the American continent. In 1824 the Hudson's Bay Company established an outpost at Fort Vancouver, 40 miles downstream from Wind River on the Columbia, as a base for hunting beaver and otter. The early trappers depended on the knowledge from people of many tribal groups to help them navigate the steep, tangled terrain of the Cascade Range. People from the Yakima, Klickatat, Wishram, and Cascade groups knew this country. For thousands of years they enjoyed a thriving economy based on the seasonal bounty of its rivers, forests, and meadows. But the Hudson's Bay Company and other ventures launched from back East approached the economic value of the region's natural resources on a much larger scale than the traditional use by Native Americans. Exporting furs to China, salmon to London, and timber to Honolulu, the exploitation of natural resources far exceeded what the local population had traditionally consumed. According to historian Carlos Schwantes, the era of fur trade began a pattern of large-scale resource extraction in the Pacific Northwest that would continue well into the 20th century.[9] Competition was fierce among the British and American companies vying for a piece of the Northwest's riches. Fur trapping was a seasonal venture for the Hudson's Bay Company, so lumbering became a lucrative use of men and ships during what one company captain called "the dead season of the year." With the construction of a sawmill on the Columbia River in 1827, timber soon surpassed pelts in economic importance and became a mainstay of the northwest economy for the next 160 years.[10]

A handful of sawmills dotted the Columbia and its tributaries by the 1840s, when settlers began moving west by way of the Oregon Trail. Most homesteaders headed south from the Columbia River and toward the fertile lowlands of the Willamette Valley. By the 1880s most of these lowlands had been settled and latecomers had to venture farther up river valleys into the Cascades to stake their homestead claims. In the upper Wind River valley before the turn of the century, a few homesteaders got a toehold on land that had been partially opened by wildfires (fig. 4). Clearing burned stumps and snags was backbreaking work, and even after the land was cleared, agriculture proved to be difficult in the Wind River valley. The rugged

Fig. 4.
Homestead of
Elias Wigal,
district ranger,
1904–1911,
in the upper
Wind River
valley, 1891.
(Courtesy of
USDA Forest
Service.)

countryside could drain cold air and bring unexpected frosts almost anytime of year. To supplement farming, most homesteaders looked for additional employment in the forest. Many worked at a steam-powered sawmill that operated for a short time in the mid-1880s in Carson, downstream on Wind River toward the Columbia. But the mill was transient, like the timber industry itself, and within a few years it closed down. Settlers were transient too. Many came to Wind River only to find that farming was too difficult and soon sold their holdings to timber companies that were sprouting across the isolated territory north of Oregon and the Columbia River.[11]

Washington's statehood in 1889 ended that isolation. In the next few decades, Washington's population burgeoned as railroads opened the new state to the rest of the nation and the world. Trains carried timber east and returned with loads of new residents. Ships carried Douglas fir timber from the forests of Washington to population centers throughout the Pacific, where they were used to build the wharves of San Francisco and Honolulu and the mines of Chile, Peru, and Australia.[12] The young state of Washington rapidly established itself as the number-one timber producer in the nation, with Douglas fir its main export.[13]

Throughout most of the 19th century, lumbering in the United States was an industry of boom and bust. Following the Civil War, New York had been the largest timber producer in the nation. As its forests were depleted, lumbering moved on to the Great Lakes region, where vast stands of white pine fed the demand of an expanding nation. Then as timber supply diminished in the Lake States, speculators in the 1890s began to look to the Northwest for timber. In 1902, the year of the Yacolt fire, Henry Gannett of the U.S. Geological Survey published an assessment of the Northwest forests, in which he wrote that "the forests of western Washington are among the densest, heaviest, and most continuous in the United States. Except for a few prairie openings, and except where removed by fire and the ax, they cover the country as a thick mantle from a line high up on the Cascade Range westward to the Pacific."[14]

Timber harvests in the Cascades progressed slowly, as ox teams trudged logs out of the woods on skid roads. But by the end of the 19th century, new technologies began to speed the rate of logging and extended the reach of the logger. Steam-powered winches, called donkeys, replaced ox teams and skid roads. Logging railroads snaked into forestlands, providing access beyond the waterways. Overhead cables, strung from ridge to ridge in narrow valleys, made it possible to reach timber in the steepest terrain and drag logs down mountain slopes. And the new Northern Pacific and Great Northern railroads funneled wood from the Pacific Northwest to lumber-hungry markets beyond the region.

Speculators saw their opportunity in the backwoods homesteaders of southern Washington. Land title records from Skamania County document a rush of homestead claims within the Wind River valley in 1897 and 1898.[15] Virtually worthless as farmland, the claims were held only long enough to transfer ownership to so-called locators who recognized the land's timber value. Many of these holdings consolidated into the Wind River Lumber Company around the turn of the century. According to historian Margaret Felt, the Wind River Lumber Company eventually "owned, at one time, more than one billion board feet* of timber 200 years of age or over in the Wind River area alone."[16]

During its first few years, the Wind River Lumber Company restricted its logging to forest stands close to Wind River, where loggers could drive their ox teams and build ponds to store logs. The lumber company built splash dams on Wind River and its tributaries, Trout and Panther

---

* A board foot is a volume of wood measuring 1 foot by 1 foot by 1 inch.

Fig. 5. Splash dam on Trout Creek, ca. 1910. (Courtesy of USDA Forest Service.)

creeks, which held back enough water to hold several hundred logs at a time (fig. 5). The company used the dams to store logs and the river itself to deliver them to markets downstream in tumultuous river drives. To power the river drives, they opened floodgates of the upper splash dams and swept tons of logs downstream. Along the way, lower floodgates opened to push the churning mass of timber on toward the Columbia River. Drivers armed with long poles prodded the logs through tight spots and kept the drive moving until it was delivered in a choking flood to the Columbia, where it was then rafted across the river to the lumber mill.

Within a few years the trees closest to the river had been cut and the lumber company began to build logging railroads to reach timber farther from the waterway. Eventually the Wind River Lumber Company built 30 miles of main railroad line and numerous spurs throughout the upper valley. The new transportation technology increased the volume of timber taken from the forest and increased the amount of wreckage left behind. Steam donkeys dragged heavy logs across the ground, leaving behind a tangle of broken trees, branches, and stumps (fig. 6). Sparks flew from the trains' brakes and smokestacks, igniting the slash. Even the cigars of smoking passengers could set fire to tinder-dry debris that lined the tracks. Often the fires that ignited dry, charred wood burned much hotter than fires in standing green forests with damp understories.

Fig. 6. Forest Ranger Fritz Sethe (center) inspecting Wind River Lumber Company steam donkey, ca. 1913. (Courtesy of USDA Forest Service.)

For years, slash fires were a common hazard to logging operations and were treated as part of the cost of doing business. With the big trees taken out and the leftovers in a dry heap, lumber companies moved on to new territory rather than bother to clean up and replant the forest. In Washington, D.C., at the turn of the century, politicians and government officials saw wildfires and the wasteful practices of cut-and-run logging as double threats to the nation's forests. Their concern went all the way to the president of the United States. In an attempt to protect the remaining unclaimed forests, legislation in 1891 gave presidential authority to designate forest reserves within the public domain. In 1893 President Harrison established the Pacific Forest Reserve, which included the Wind River valley. In 1897 President Cleveland enlarged and renamed the protected land the Mount Rainier Forest Reserve. Much of the nation outside the West saw the reserves as a laudable attempt to take back the public domain for the good of the public. Some westerners saw it as a land grab. Opposition was especially strong in Skamania County, where 80 percent of the land lay within the new reserve boundary.[17]

Additional legislation in 1897 assuaged some industry fears by clarifying that the purpose of the federal forests was to protect the nation's water and provide a steady supply of timber to meet the needs of American citizens. The purpose of the forest reserves, therefore, was foremost economic, as historian William Robbins pointed out.[18] Establishing forest reserves reversed a trend in national policy, which had been in place for most of the 19th century, of disposing federal land to nonfederal owners. The forest reserves retained federal ownership and focused on federal management on these public lands. With the 1897 legislation, the General Land Office received authority to hire seasonal rangers to administer the forest reserves.

In 1902 Horace Wetherell, a homesteader from the Wind River valley, was appointed as a forest ranger to patrol the rugged region of federal forest between Mount St. Helens and the Columbia River Gorge. His job was to post boundary notices, survey homestead claims, prevent timber theft, and put out fires in the southern part of the Mount Rainier Forest Reserve. Although Wetherell was a local farmer, most settlers considered forest rangers agents of an alien policy. Charles Cowan, a fire patrol ranger who began his career a few years after Wetherell, recalled a feeling of antagonism among the locals, that the rangers were preventing settlers from a legitimate function of land clearing. Charges of setting illegal fires brought against settlers by rangers often got little support in court, because laws were aimed at people who deliberately set fire on someone else's land with the intent to do harm. "Now, the intent to do harm," wrote Cowan, "was where the sticker came as far as the local magistrate was concerned. There was certainly no harm in burning up timber that nobody was using . . . They couldn't prove neglect because [the settler] hadn't neglected anything in setting the fire."[19]

Wetherell patrolled hundreds of miles without trails, buildings, or phone lines. He was required to furnish his own horse, saddle, and any other equipment to carry on his work, for which he was paid $60 a month.[20] Within three months, he was fighting the biggest fire of his life.

Chapter One
State of the Science in 1908

⚹

*Nowhere is the record of experience and the historic method
of study of more value than in an empirical art like forestry,
in which it takes decades, a lifetime, nay a century to see the
final effects of operations.*

—Bernhard E. Fernow[1]

In 1873 the American Association for the Advancement of Science
convened in Portland, Oregon. Among the hot new science topics of
the time was the relationship between forests and water, inspired by
naturalist George Perkins Marsh's book *Man and Nature*. Marsh had
fired the nation to thinking about the increasing possibility of floods,
drought, and a nationwide timber famine due to the unchecked
harvest of the nation's forests. In response, the scientists convening
in Portland recommended that Congress take action to promote "the
cultivation of timber and the preservation of forests." Three years later,
Congress directed the Secretary of Agriculture to "appoint a man of
approved attainments and practically well acquainted with the methods
of statistical inquiry" to assess the nation's supply of timber and other
forest products. The job description was exhaustive, including the
calculation of "the probable supply of future wants, the means best
adapted for the preservation and renewal of forests, the influence of
forests on climate and the measures that have been successfully applied
in foreign countries or that may be deemed applicable in this country
for the preservation or planting of forests."[2]

The man they appointed to assess the nation's timber supply was
Franklin Hough, a New York physician self-taught in natural history
and forestry. Despite his lack of formal training and the enormity of

the job, Hough managed to document what was known at the time about the nation's forests. To learn more about the state of the science, Hough traveled to Europe and toured forestry schools in Germany, France, and Prussia. Impressed by European experimental forests expressly designated for research, Hough recommended a similar American system of forest experiment stations and European-trained foresters to manage the nation's forestlands. In the 1870s, however, Hough's suggestion fell on deaf ears, and his job was passed on to another self-taught naturalist, Nathaniel Eggleston, in 1883 and then, in a step toward realizing Hough's recommendations, to Bernhard Fernow in 1886 (fig. 1.1).

Fernow, a Prussian immigrant, had trained as a forester at a time when European forestry was becoming more public, scientific, and experimental. Before the middle of the 18th century, European forests were generally private estates, and the "forest service" was primarily responsible for providing wildlife for the nobility's feasting and hunting pleasure. But as industrialization pushed up wood prices, the need arose for timber management, and a handful of scientists emerged early in the 19th century to define a new science of forestry in Europe. Among those first European forest scientists was Heinrich von Cotta.

Fig. 1.1. Bernhard E. Fernow, ca. 1895. (Courtesy of the Forest History Society.)

An accomplished plant physiologist, Cotta founded the Royal Saxonian Forest Academy in the early 1800s, the first forestry school in the world. He was among the first to scientifically investigate changing forests, examine the decline of the European forest, and consider ways to regenerate it. In 1816 Cotta published one of the first forestry texts, *Advice on Silviculture*. Although probably too young to have ever met Cotta personally, Fernow admired the German master and studied his forestry text. So moved by the cautionary tone of the book's introduction, Fernow translated the preface to Cotta's text in 1902 and published it in the first issue of *Forest Quarterly*.*

Cotta described two different approaches to studying forests, embodied by the "empiricists" and the "scientists." This division would continue as empirical management and research science evolved over the next 200 years. "Rarely are both united," Cotta wrote. He reminded the empiricists that the forest is far more complicated than can be measured. And he reminded forest scientists that the purpose of good science is not a search for answers as much as learning to ask the right questions. According to Cotta, those questions revolved around three challenges: "first, the long time which wood needs for its development; second, the great variety of sites on which it grows; thirdly, the fact that the forester who practices much writes but little, and he who writes much practices but little." Cotta wrote that the long time it takes for a forest to develop causes some forest practices to be considered good for a time but they are then seen as detrimental later. Similarly, the great variety of forest conditions causes some practices to be successful in one place while failing in others. He wrote, "And the third fact brings it about that the best experiences die with the man who made them, and that many entirely one-sided experiences are copied as articles of faith which nobody dares to gainsay."[3]

Thanks to Cotta and his peers, German forestry became the standard of the science by the mid-1800s. However, German forests had been harvested for centuries and were nothing like the forests on

* *Forest Quarterly* was one of the earliest forestry journals in the United States. Fernow was editor from 1903 to 1916. A year later the journal merged with the Proceedings of the Society of American Foresters to become the *Journal of Forestry*.

the American continent, especially those in the Douglas fir region, which were big, old, and messy. In the early 1800s German foresters had observed that pure stands of spruce produced more timber than a forest of mixed species. They set out to convert all their forests to pure stands by clearcutting the mixed forests and replacing them with spruce plantations. Many decades later, they saw yields diminish from subsequent planting on exhausted forest soils. By Fernow's time, German foresters were abandoning pure plantations and returning to mixed forests.[4] Fernow immigrated to the United States in 1876, four years after a system of experimental forests had been established in Germany. As head of the Bureau of Forestry within the Department of Agriculture, Fernow made every attempt to infuse what he could of European science into the management of American forests. He spoke to groups as diverse as charcoal workers, railroad-tie manufacturers, and the American Forestry Congress, outlining 12 principles of silviculture. Among them was the importance of having a mix of tree species and age classes. Fernow wrote, "The mixed forest affords greater security against damages by wind, fire, frost, snow, diseases, besides yielding a larger amount of wood."[5]

Above all, Fernow recognized that to apply scientific management to American forests, he would have to show that such management was profitable. For his audience—industry landowners, government agencies, and newly emerging conservationists—Fernow defined forestry as two branches of production: "one branch being concerned with the production of the material, the other with production of revenue."[6] To underscore his economic frame of mind, Fernow tabulated the cost *per page* of the scientific findings he reported in his *Report upon the Forestry Investigations of the USDA. 1877-1898*. He valued his findings at an estimated $75 per page in this 401-page report.

But perhaps to keep his expenses down, Fernow had not seen firsthand some of the forests he reported upon. As a result of not taking what he referred to as "an unnecessary junket" out West,[7] Fernow's report did not mention the Cascade Mountains. He detailed the productivity of the "ponderous bull pine" in the intermountain west but hardly mentioned Douglas fir. And despite his efforts to reach various audiences, Fernow held little sway with public opinion. Throughout his 12-year tenure as

Fig. 1.2. Gifford Pinchot, ca. 1900. (Courtesy of USDA Forest Service, Grey Towers.)

chief forester, his agency was chronically understaffed and vulnerable to budget cuts. He saw that "fickle public opinion may subvert at any time well laid plans which take time in maturing."[8] Frustrated, Fernow resigned to organize the nation's first forestry school, the New York State College of Forestry.[9]

The new man who took over the Bureau of Forestry in 1898 was Gifford Pinchot (fig. 1.2). Fernow described his successor as "young, ambitious, aggressive, with some knowledge of forestry acquired in Europe and with influential connections and a large fortune, he easily procured the first need for effective sowing on the well-plowed field before him—appropriations."[10] Pinchot distanced himself from Fernow soon after taking office. Unlike Fernow, an immigrant and a political outsider, Pinchot was a member of the rich and powerful class that ran the nation at the turn of century. He had the political connections to secure funds and the personal charisma to inspire supporters, and he saw the scientific management of the nation's forest reserves as a calling to both political power and public service. Pinchot embraced the ideals of the Progressive era, a time in the early 20th century when leaders believed that a systematic, scientific approach to the natural world would bring the greatest good to the greatest number of people. As historian William Robbins has argued, the Progressives valued nature for its commodities for human use, a reality that was especially apparent in the resource-rich Pacific Northwest.[11] According to historian Richard Rajala, Pinchot's model of technological efficiency drew in part from industrialization of the late 19th century that strove to mechanize all aspects of nature.[12] Pinchot deplored the waste he saw in the forests across the nation. As chief forester, he wanted to clean up all that was corrupt, unorganized, or wasteful in the nation's forests.

At the time Pinchot took office, America's vast public forest reserves were administered by the Government Land Office (GLO) in the Department of Interior. Pinchot's forest research and management recommendations were coming from the landless Bureau of Forestry within the Department of Agriculture, a bureaucratic world away. With a mere token of appropriations from Congress, the meager, untrained GLO staff could not begin to manage the vast federal forests of the West. Increasingly the GLO and Congress turned to the bureau foresters for advice.

Although Pinchot emphasized practical aspects of forestry, research was an essential part of Pinchot's Bureau of Forestry. He established a Section of Special Investigations in 1898 to gather the information necessary for science-based management of the forests. The section grew into the Division of Forest Investigation by 1902 and claimed one-third of the bureau's budget. Pinchot added a Section of Silvics in 1903, headed by Raphael Zon, an Eastern European immigrant whom Bernhard Fernow had mentored.

Pinchot's bureau within the Department of Agriculture continued to grow under the Roosevelt administration, as did his frustration with secondhand control over the nation's forests. Finally, in a proclamation written by Pinchot himself, President Roosevelt transferred the administration of 63 million acres of the nation's forests from the Interior Department to Pinchot's Bureau of Forestry in 1905. The agency was renamed the Forest Service to reflect Pinchot's philosophy of public service.[13]

Pinchot saw an opportunity to manage the forest reserves sustainably, as a crop that is planted, harvested, and cultivated by skilled experts for the greatest benefit to the greatest number for the longest time. The greatest good for the greatest number may have been a concept borrowed from British philosopher Jeremy Bentham, according to Pinchot biographer Char Miller. According to Miller, the addition of the idea of the longest time committed Pinchot's agency to "the creation of sustainable economics and ecosystems, a potentially important notion in a society so wedded to profligacy and waste."[14] Sustainability was a new way of thinking for a nation that had just pushed its way to the edge of its seemingly endless frontier.

"The purpose of forestry," Pinchot wrote in his autobiography, "is to make the forest produce the largest possible amount of whatever crop or service will be most useful, and keep on producing it for generation after generation of trees and men."[15] Over time, he marshaled a corps of disciples that reflected both his sense of urgency about sustainable use and his sense of optimism that both sustainability and use could be achieved. These disciples included forest rangers, who served on the front lines of the cause, and inspectors, who provided a scientific basis for the cause. Rangers, such as Horace Wetherell from the Wind River area of the Mount Rainier Forest Reserve, were drawn mostly from the ranks of local men. Forest inspectors were more often new graduates from one of the handful of forestry programs springing up at American universities. An exception was a forest inspector in the Douglas fir region named Edward T. Allen.

Unlike most of the scientific men recruited into Pinchot's army, Allen was no stranger to the Northwest woods. Born in New Haven, Connecticut, he had been tutored at home by his father, a former professor at Yale. His family packed off for the West when Allen was a child, and by 1889, at age 14, "he was living with his father in a wilderness home at the foot of Mount Rainier, accessible only by foot or horseback."[16] As a newspaper reporter, Allen witnessed firsthand the changes swirling around the Pacific Northwest at the end of the 19th century. Wealthy lumbermen organized in associations while their laborers organized into unions and the federal government organized its Forest Service. Allen returned to the woods in 1898, eventually working for the Department of Agriculture's Bureau of

Forestry. In 1903 he produced the 94-page report titled "Red Fir in the Northwest," the nation's first assessment of one of its important commercial tree species, the Douglas fir.

Allen's observations were careful and thorough. He noted that "few [other tree species] promise to exert such an influence on the lumber supply of the future." He recorded how Douglas fir seeds sprouted only where the land was clear of forest litter, and how they succeeded only where fire prevented western hemlock from establishing. He measured rate of growth from annual rings and saw that Douglas fir growth peaked at about 80 years. From his observations, Allen concluded that Douglas fir required clearcutting to flourish and that the "greatest production of wood can be secured by cutting every 90 years." Subsequent research would confirm many of Allen's conclusions about Douglas fir.[17]

Allen later served as California's first state forester and had a hand in early management of the forest reserves in Oregon and Washington, helping to draw up one of the region's first timber sales.[18] He helped Pinchot outline the extended boundaries of the new national forests. From that work in 1907, the Columbia National Forest, including Wind River and much of the Yacolt Burn, was carved from the southern portion of the Mount Rainier Forest Reserve. The next year, when Pinchot shuffled the national forests into six inspection districts (which were redesignated as regions in 1930), he named Allen as the first forester for District Six, which at the time included Washington, Oregon, and Alaska.[19] The district headquarters was located in Portland, with branches specializing in silviculture, grazing, forest products, and planting. Allen quit the agency in 1909 to head the Western Forestry and Conservation Association, one of the new timber industry associations forming in the West.[20] As one of a number of industry spokesmen, he became especially effective in helping to secure congressional appropriations for fire protection and forest research in the Pacific Northwest.

Rebounding from an economic depression in the 1890s, the Pacific Northwest was emerging as an economic engine, no longer a backwater to the rest of the nation. In the first decade of the 20th century, Washington's population doubled and Oregon's population increased by two-thirds.[21] In 1905 Portland hosted the Lewis and Clark

Centennial Exposition to showcase the region's economic potential in the coming century. One feature in particular, the Timber Palace, symbolized the region's timber bounty. Made entirely of old-growth Douglas fir logs, the palace stood 100 feet wide, 200 feet long, and 72 feet tall. It introduced a new product called plywood, and showcased the grandeur and promise of the Northwest forests. Boosters hoped that the Timber Palace would define the Portland exhibition just as the Eiffel Tower had defined the Paris Exhibition five years earlier.

And so the 20th century began with grand expectations for science, technology, and forests in the Pacific Northwest.

Chapter Two
Early Science in the Douglas Fir Region

❧

By and large, we greenhorn foresters from the eastern forest schools got accepted quite quickly.

—Thornton T. Munger[1]

Although Gifford Pinchot had been trained in Europe as a forester, he was reluctant to import European foresters to staff his new agency. In 1900, at Pinchot's suggestion, his family endowed a school of forestry at Yale University, with the objective of producing "American foresters trained by Americans in American ways for the work ahead in American forests."[2] Some of the first forest scientists to go to Wind River were fresh from Yale, armed with their experience from the East Coast and Europe. They discovered new challenges in the unruly, wild nature of the Pacific Northwest. Among the stands of giant, ancient trees, they saw places denuded by fire and careless harvest. Privately owned forests were abandoned after harvest, left to the whims of encroaching brush and frequent fire. Even the national forests were ignored following harvest.

The first order of business for the new forest researchers was to survey the forests. Maps of the Mount Rainier Forest Reserve, published in 1900 as part of Henry Gannett's work for the U.S. Geological Survey, included general descriptions of forest types and condition, hydrology, fire history, and grazing use. But few records were kept following timber sales, and little was known about the land's ability to regenerate. This was all about to change.

"The practice of forestry is still young in this district," wrote Fred Ames, the first head of silviculture in District Six, "and we have much to learn regarding the practical management of forests."[4] Beginning in

1908, Ames directed an inventory of forests in the Pacific Northwest. He asked for a full accounting of standing forests and cutover areas, detailed enough to answer future questions. He hired crews of college men to spend their summers in the woods as estimators, compassmen, and draftsmen, recording tree species and age, estimating timber volume, and assessing the number of trees they considered overmature, defective, or lacking in marketable value. Scientific crews often shared the tasks of packer and cook. One of those crew members was T. J. Starker, who was among the first graduates of Oregon Agricultural College's forestry program, which was established in Corvallis in 1906.

"Going back to 1909, I spent a summer on the Columbia National Forest mapping the timber types," Starker recalled years later. "There were no maps so we climbed the higher elevations and mapped in the age groups as best we could. Called 'Extensive Reconnaissance.'"[5]

Another young researcher in District Six was Thornton T. Munger (fig. 2.1). Following graduation from Yale and a three-month stint working with Raphael Zon in the Washington, D.C., office of the Forest Service, T. T. Munger headed west in 1908. The young New Englander had been assigned to study the encroachment of lodgepole pine into ponderosa pine on the eastern slopes of Oregon's Cascade Mountains. Munger may have hoped for a more romantic assignment. "Some people looked down on research in those days and would have considered [this assignment] a lemon," he remembered later. "In general the more romantic administrative frontier work appealed to them more than what they considered the ring-counting work of a researcher."[6] But he took on the task, spending three months in the saddle, riding hundreds of miles and recording his observations throughout the pine forests of eastern Oregon. "I was very fortunate when I went into this central Oregon country that I could ride horseback, because if you didn't know how to handle a horse, you didn't get any respect from the westerners in those days."[7]

When his work was completed, Munger stayed at District Six as a one-man research unit, the Section of Silvics within the branch of silviculture. In 1908 silvics was the study of habits and the natural history of forest trees and the basis for all practical silvicultural operations, as defined by George Sudworth, the Forest Service

Fig. 2.1. Thornton T. Munger, 1924. (Courtesy of USDA Forest Service.)

dendrologist who wrote one of the first compendiums of forest trees on the Pacific slope.[8] One of Munger's assignments during his first winter in the Pacific Northwest came directly from the secretary of agriculture James Wilson, who thought the Forest Service was too slow in replanting burned forestlands. According to Munger, it did not matter to Secretary Wilson that the Forest Service had nothing to plant—no seedlings nor even any seeds for planting. "That doesn't make any difference," Munger recalled Wilson saying, "get some seed from Europe, or get some seed from the East."[9]

To carry out Wilson's directive, Munger and a crew rode up the flanks of Mount Hood in the Oregon Cascades, armed with several sacks of tree seed from Europe and the East Coast. "In the night our tents began to sag and when we got up we found there was six or eight inches of snow on the ground, and it was snowing hard." The men had to get out quickly. They jettisoned the seed onto the snow and beat a hasty retreat down the mountain. "People who have gone through that area since then are surprised to discover once in a while an eastern oak or a European pine of some kind, and wondered how in the world it got there," Munger recalled many years later.[10]

By the spring of 1909, Munger had a more rational and progressive research plan in mind. He focused not on eastern oaks and European pines but on the dominant native trees of the region, particularly Douglas fir. For much of the next four decades, Munger would continue to conduct forest research from the Portland office, helping to define what was known about Northwest forests. That office would become his pulpit, where Munger, the son of a New England minister, would work to convert the wasteful practices of Northwest timbermen and redeem the ravaged forest.

Munger seemed to share the determined purpose of his Yankee preacher father. The elder Munger was described as speaking "wisely and shrewdly about thrift and health, at the same time that he speaks inspiringly about courage and purpose."[11] In time, the younger Munger would speak wisely and shrewdly about the thrift and health of forests and inspiringly about the purpose of experimental forests.

"From the start, I was not interested in research for research's sake but wanted to see research put into use," Munger recalled.[12] The scientific research he conducted was a pragmatic response to immediate problems he saw in the Pacific Northwest forests. He wanted to understand how to grow trees as a continuous crop and find ways to manage the forest for sustainable wood production and sustainable profits. It was science in the service of society, reflecting a society that valued efficiency, thrift, and productivity.

Turning his attention to the tree that dominated the western slopes of the Cascades, Munger found a natural laboratory in the Wind River area for studying Douglas fir. Fires had left a patchwork of even-aged stands that were just right for comparative study. "We studied stands all the way from 30 years old up to 125, and they weren't very hard to find because there was a great deal of second growth Douglas fir on old burns."[13] In the burned hillsides of the western Cascades, Munger set up sample plots within each of the age classes of Douglas fir and on different qualities of land. He measured 40-year-old trees and compared them to the size of 50-, 60-, 70-, and 80-year-old trees. He compared trees on similar sites so that just the growth and not the favorability of the site would be measured.

Only a few months were dry enough in the western Cascades to allow such fieldwork. During the long, wet winter, he employed "computers," which in 1909 were men hired to compute numbers into tables to compare growth rates. Munger pored through the computed growth tables and saw how the numbers increased rapidly across the age classes. Munger's documentation showed that Douglas fir grew very fast, as E. T. Allen had observed several years earlier. Like Allen, Munger observed how the Pacific Northwest's timber industry mined trees in the same pattern of cut-and-run harvest that had wasted so much forestland in other parts of the country.[14] Munger's plots provided evidence that reforestation could be profitable. He, like Allen before him, believed that Douglas fir, more than any other species in the Northwest forests, could be managed in a way to regenerate successive timber crops that could sustain the lumber industry and provide for local communities. The forest management Munger envisioned would mimic the way fire had established blocks of young trees that were all the same age and all the same size and that would eventually be harvested all at the same time.

Munger had been working in the Douglas fir region for only two years when he published his 27-page report, *Growth and Management of Douglas Fir*.[15] Much of what he wrote was consistent with the much longer treatise Allen had written just eight years earlier. But Munger neither referenced Allen's work, nor did he credit Allen. When asked about Allen's influence, Munger replied, "Ten years before, E. T. Allen . . . made some study which was never published but was available in manuscript form. Our work did not hinge on that study at all."[16]

Nonetheless, Munger's report was prescient. In it, he predicted that "the stands of the future can be made to produce more and better timber to the acre than have the average forests of the past." Foreseeing a depletion in the future timber supply, he predicted that "within three or four decades the demand for even 50-year-old Douglas fir will be very keen."[17] He recommended methods of scientific management that would create forests that he believed would be even more productive than the wild forests they replaced.

But Munger needed evidence if he was to persuade forest owners. For that purpose, in 1910 he established acre-size plots in even-aged young stands on old burns in western Oregon and Washington (fig. 2.2). He tagged and recorded the size of every tree within every plot,

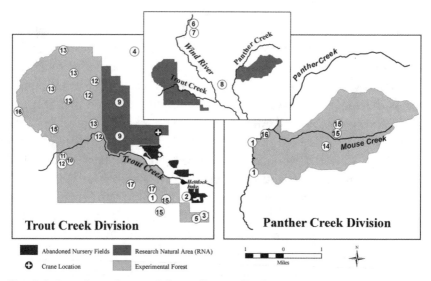

Fig. 2.2. Locations of some of the studies at Wind River Experimental Forest and its environs. Not all studies described in the text are included here, because specific locations are not known or permanent markers do not exist. The names and dates refer to the scientist(s) who initiated the study and the year the study began. Many study sites were used by more than one scientist; these additional studies are referred to in the text. (Map by Theresa Valentine, USDA Forest Service, PNW Research Station.)

1. *Permanent Growth Plots—T. T. Munger, 1910*
2. *Wind River Arboretum—Munger, 1912*
3. *Heredity Study—Munger, 1915*
4. *Wind River Transect—Hofmann, 1917*
5. *Precommercial Thinning—Hofmann, 1919*
6. *Spacing Study—Leo Isaac, 1925*
7. *Regional Races of Ponderosa Pine—Munger, 1926*
8. *Crop Tree Pruning—Civilian Conservation Corps (CCC), 1935*
9. *Research Natural Area Permanent Plots—Munger, 1947*
10. *Thinning Study—George Staebler, 1953*
11. *Red Alder–Douglas Fir Study—Robert Tarrant, 1956*
12. *Fertilization and Spacing Studies—Richard Miller and Tarrant, 1964*
13. *Trout Creek Hill Spacing and Species Trials—Dean DeBell and Constance Harrington, 1979*
14. *Shelterwood Cutting—Dean DeBell and Jerry Franklin, 1986*
15. *Gap Study—Thomas Spies, 1990*
16. *Riparian Permanent Plots—David Shaw, 2003*
17. *Additional permanent plots added to 1910 Munger Permanent Growth Plots, 2004*

and planned to remeasure each of the trees in his permanent plots every few years in perpetuity. Munger was 26 years old at the time. He knew that the data he needed would accumulate slowly, but that long-term field observation was the only way to understand forest changes that would occur over time. Remeasurement of these plots continued throughout Munger's career and well beyond. Over time, some were logged or blown down and salvaged, but the plots that remain are still remeasured in the 21st century. Munger believed that the permanent plots offered the most convincing evidence that lumbermen needed to show that trees grow at a predictable rate. "We pioneered in this field of permanent growth plots in the West," he recalled years later (fig. 2.3).[18]

Munger strove to show that reforestation could be profitable, yet reforestation was not possible without a ready source of seeds or seedlings. Based on observations he and others had made of the

Fig. 2.3. Thornton T. Munger remeasuring the diameter of Douglas fir in one of his permanent growth plots, 1939. (Courtesy of USDA Forest Service.)

growth habits of Douglas fir, Munger advocated that the forest should be clearcut with all merchantable timber removed and with care taken to leave some standing trees nearby to provide seed for future reproduction. He recommended burning the remaining vegetation and debris left over from logging to expose the mineral soil below the forest litter. With soil exposed for a seedbed and a ready source of seed close by, Munger was confident that a new timber crop could take hold if the land were protected from subsequent fire.[19]

So as early as 1911, Munger had identified Douglas fir as the preferred forest crop that would be managed with clearcutting and slash burning to guarantee reproduction as quickly as possible. These methods would become the tenets for most forest management in the Pacific Northwest for the next 75 years.

While Munger mused about growing better trees in the forest, Julius Kummel planned how to grow trees in a nursery. Kummel, another young forest researcher working under Fred Ames, headed the District Six Section of Planting with the charge to develop a source of seedlings to replant the land that had been left bare by recent forest fires. Notwithstanding Munger's experience on Mount Hood, attempts had failed to regenerate forests using broadcasted seed. Of particular concern was the Bull Run watershed. Bull Run supplied water from the slopes of Mount Hood to the growing metropolis of

Fig. 2.4. Elias Wigal constructing the first ranger station at Hemlock, 1906. (Courtesy of USDA Forest Service.)

Portland. It was one of the original forest reserves designated in 1891 in recognition of its importance as a water source for the bustling city. Two fires in Bull Run, in 1873 and 1881, had burned over 7,000 acres and increased erosion on the slopes and sedimentation in the streams. The forest there needed to be replanted.[20]

The Forest Service needed seedlings—lots of seedlings. Kummel estimated that it would take an annual production of two million tree seedlings to replant the forests of the region and keep them stocked after future harvests.[21] Kummel drew up a plan and chose a site at Wind River as a good place for a tree nursery, where the ground was flat and the rainfall abundant. However, two other features may have been considered strong advantages. First was the one-room Hemlock Ranger Station on the south side of Trout Creek. The second was Elias Wigal, who had replaced Horace Wetherell in 1904 as district ranger and who had built the ranger station building (fig. 2.4). From this new station, located across from the Wind River Lumber Company, Wigal had administered the district's first timber sale in 1906.

The work to clear five acres for the nursery would be backbreaking. Though the site was level, it contained huge stumps and snags that had withstood successive fires and logging. Wigal would need help. His journal entry for November 26, 1909, reads, "Snowing; left home; went to Carson to get men to look over ground and take contract to grub five acres on hemlock [sic] Station that mr. [sic] Kummel wanted to use for nursery."[22] Over the winter, Wigal hired three men and began breaking new ground for the nursery.

Wigal and his crew exploded the big stumps with blasting powder. Then they hacked out large blocks of wood, split them into chunks, and piled the chunks around the remaining stump for burning. They repeatedly blasted, hacked, and burned the largest of the stumps until there was just a hole where the old-growth roots had been splayed. Then they filled the holes and plowed the land to dislodge smaller roots. Finally they leveled the land with a scraper, harrow, and float (fig. 2.5).

By 1910 Wigal and his crew had grubbed out five acres for Kummel's nursery, and assistant forest ranger Charles Miner was placed in charge of nursery operations. As the nursery expanded, so did the need for labor. Some workers were drawn from the ranks of the Columbia

Fig. 2.5. The Wind River Nursery fields, plowed, seeded, and covered after trees were cut and stumps blasted, ca. 1910. (Courtesy of USDA Forest Service.)

National Forest, including young Starker, who recalled grubbing out stumps with a capstan powered by a team of horses.[23] Nursery officials advertised as far away as Portland and Vancouver for seasonal laborers, offering less than commodious accommodations. Miner instructed a prospective laborer, "You should take the 8:20 morning train on the North Bank railroad . . . and go to Carson, Washington. You will probably have to walk to the nursery, which is about 12 miles from Carson. The government team will probably be in town and take your bedroll if the wagon is not overloaded."[24]

From the beginning, much of the nursery work was experimental. As Kummel had directed, Douglas fir and western white pine were the primary species grown that first year. But Kummel had also gathered seeds from around the world to test at Wind River. The nursery records document experimental plantings of eastern hardwoods, such as black walnut, shagbark hickory, white ash, red oak, and several European conifers, including Scotch pine, Norway spruce, and European larch. Much of the nursery work was done flying by the seats of their pants. Miner and Wigal built contraptions to drill a million seeds into seedbeds. Starker later recalled the makeshift technology: "In the early days transplanting was done by use of an 8-inch board with notches in which to hang seedlings at the length of the width of the bed. A trench was dug by means of a flat-ended spade and after the filled board was properly spaced the trench was filled. Then the process was repeated again and again . . . In the old days, they had a dozen or more women pulling out the weeds by hand."[25]

The Wind River site offered more challenges than the nurserymen may have bargained for. When they saw their labors being plucked from the ground by birds and mice, nursery workers doused the seed with coal tar, inadvertently ruining the seed. They struggled against the whims of the climate, keeping seedlings moist throughout the dry summer, protected from drowning throughout the wet winter, and shielded from killing frosts that tumbled down into the valley on cold nights. "And thru [sic] the years I have often thought that there must be a better location for a nursery in some Willamette Valley [site] where two-year old seedlings could be twice as large as those raised at Wind River," Starker reflected in 1975, conceding that by that time too much had been invested to change locations.[26]

The first time personnel lived at Hemlock Station through the coldest months was during the winter of 1910–11. A memo to Miner in November 1910 reminded him to put in a good supply of firewood because he, Wigal, and forest guard Fritz Sethe would be there until January, and one of them would be there throughout the rest of the winter.[27] The record does not note which man stayed through the winter, but in early spring of 1911, Wigal resigned and returned to his nearby ranch in the Wind River valley. Sethe took his place as district ranger.

Fig. 2.6. The Wind River Experiment Station and nursery fields, 1914. The Experiment Station office is the third building from the left. (Courtesy of USDA Forest Service.)

In that year, the Wind River Nursery reported an annual production of over a million seedlings. But the production came with a host of new practical questions, such as how to promote better germination, how to thin, how to irrigate, and how to prevent tender roots from rotting. The nursery's practical questions were focused on getting seeds into seedlings. But which seeds? Which were the best trees for the Pacific Northwest? The Forest Service established an experiment station at Wind River in 1912 in conjunction with the nursery to pursue these and other research questions (fig. 2.6). Sethe oversaw the construction of a research laboratory, which would include a residence for employees and a greenhouse for seed tests. Eventually the nursery operations and nursery research would become separate from forest research at Wind River, but the two groups continued to share facilities, personnel, and strong ties throughout most of the century.

In its first few years, the Wind River Nursery tested the viability of seeds from all over the world. Some seeds simply failed to germinate, while others produced weak seedlings that soon died. But seeds of some species sprouted and grew, and these trees warranted attention.

Could this land that produced such giant conifers nurture other species with such spectacular results? The question was worth testing.

In 1912 Munger established an experimental planting of 10 trees each of 16 species from the nursery's first trials. The experiment was meant to test the suitability of exotic trees to the specific climate and conditions of this part of the Pacific Northwest. In the next few years, Munger collected more seeds to test and had them grown to seedlings in the nursery. By 1914 the experimental plantings had outgrown their corner lot. Sethe's crew grubbed several more acres of stumps for Munger's arboretum.

Once planted, the trees were given very little care besides weeding, since the purpose of the experiment was to test the growth of exotic species under natural conditions. The arboretum expanded again in 1920, and many of the original trees were moved to a field of stumps south of the nursery. No longer seedlings, many of these trees died in the transplanting, and the plan to expand the arboretum was abandoned. The surviving trees were then moved back to their original location in the arboretum, and many more died. The plan changed again to arrange the trees systematically, and in 1925 more land was cleared adjacent to the original arboretum to make room for new experiments testing imported conifers. Over the years, some species have outpaced others, some showing promise while others succumbed to disease, sudden cold, or ice damage. Today the arboretum contains less than a third of the original 150 species of trees from around the world. Just beyond the arboretum fence is a forest of native Douglas fir that has grown up alongside the experimental plantings of nonnative species. The Douglas firs tower over everything in the arboretum.

Allen, Munger, and others saw Douglas fir's potential as the timber industry standard in the Pacific Northwest. In the 20th century, Douglas firs would be exported throughout the world and used successfully to establish new forests in countries such as Scotland and New Zealand. But the world could not offer any better tree for forests in the Pacific Northwest, so researchers at Wind River turned their efforts toward understanding how to grow even better Douglas firs.

Chapter Three
A Focus on Research

*The research man must anticipate coming needs. He must
of necessity be ahead of his time. That means standing alone,
exposed to skepticism and ridicule of those who live only day to
day.*

—Gifford Pinchot[1]

Just as Heinrich von Cotta had described the different approaches to
forest research in his juxtaposition of scientists and empiricists, Gifford
Pinchot's Forest Service developed complementary, sometimes
opposing forces of research and management. Forest rangers were
his empiricists, "practical men, men of action, who could make quick
decisions and yet not stumble too much." According to Pinchot, men
of science were needed to keep the practical men from stumbling.[2]
But men of science were often at odds with practical men. Pinchot
recalled, "Their presence was often resented on the ground that they
were Eastern tenderfeet (which had nothing to do with the case),
but more commonly because of their persistent and sometimes
unreasonable habit of asking embarrassing professional questions."[3]

Raphael Zon, Pinchot's Washington office chief of silvics, personified
this unrelenting pursuit of knowledge (fig. 3.1). As a student in Russia
during the turbulent years prior to the Bolshevik revolution, Zon had
been imprisoned for radical activity and organizing labor unions. He
escaped incarceration and eventually made his way to the United
States, where he studied forestry under Bernhard Fernow. Zon began
work in forest research for the Bureau of Forestry in Washington,
D.C., in 1901. Like Pinchot, and Fernow before him, Zon turned
to European traditions for a model of forest science in America. He

claimed that the "heroic phase of the conquest of the West" was over, and it was time for trained researchers to confront forestry problems with objective scientific information.[4] He recommended separating Forest Service research from administration, a strategy the agency implemented in 1915 when it established the Branch of Research.[5]

With Pinchot's support, Zon traveled back to Europe in 1908 to examine the forest experiment stations that Fernow and Franklin Hough had long advocated for the United States.[6] Inspired by what he saw, Zon drew up a plan modeled after the European system, with forested areas representing each forest district across the country. These areas would be designated for "experiment and studies leading to a full and exact knowledge of American silviculture, to the most economic utilization of the products of the forest, and to the fuller appreciation of the indirect benefits of the forest."[7] For the first forest experiment station in the United States, Zon chose Fort Valley in the ponderosa pine forests of northern Arizona.* Zon reportedly claimed in 1908, "Here we shall plant the tree of research."[8]

The tree flourished. Within the next few years, the Forest Service established experiment stations in other parts of the West and began searching for an appropriate site in the Douglas fir region of the Pacific Northwest. The record is not clear if there were any other serious candidates, but Wind River seemed the natural choice: the Hemlock Ranger Station was expanding, the nursery was growing, and several research studies were underway. As director of silvics, T. T. Munger understood the value of a forest experiment station at Wind River. But Munger did not have to import a European-trained forester to direct the proposed Wind River Forest Experiment Station. In midwinter 1913 Munger traveled to Idaho to meet Julius Valentine Hofmann and size him up for the job of director (fig. 3.2).[9]

Hofmann presented impeccable credentials. A graduate of the University of Minnesota school of forestry, he was the picture of the new professional American forester. Described as an entrepreneur in a new science and a new agency, Hofmann was at the time a forest

* Fort Valley, Wind River, and most of the early experiment stations eventually became experimental forests as the stations moved their headquarters away from field settings to more centralized and typically more urban centers.

assistant at the Priest River Forest Experiment Station in Idaho. With a special interest in the natural regeneration of forests, Hofmann was well versed in the new science of ecology. Following Munger's visit, Hofmann transferred to Wind River, and on July 1, 1913, the Wind River Forest Experiment Station was officially established and Hofmann was named its director.[10] Hofmann's entrepreneurial approach would eventually clash with Munger's conservative and thrifty manner. But with everything to learn about the forest, the two scientists worked together for a decade. Their foremost concern was the regeneration of burned and cutover land.

One of Hofmann's many research interests was the physics of fire, including an analysis of weather conditions that led to forest fires. The Yacolt fire was just one of many large conflagrations in the Douglas fir region in the early 20th century. Unpredictable and deadly, forest fires were feared by timbermen. One key to controlling wildfire was to be able to predict when and where fire might strike. Hofmann worked with Bush Osborne, the forest assistant who had accompanied

Fig. 3.1. Raphael Zon, 1921. (Courtesy of USDA Forest Service.)

Fig. 3.2. Julius Hofmann, overlooking the Yacolt Burn a decade after the fire. [exact date unknown]

Munger on his snowy trip to Mount Hood, to examine the details of past fire events for clues that would help them predict the likelihood of future forest fires. Hofmann and Osborne compared fires that seemed to have had similar weather conditions and terrain and even similarly equipped firefighting crews, and yet they found that while one fire had been contained, another had burned out of control. Hofmann noticed one measure that stood out among the data: relative humidity. In time, Hofmann's study determined that relative humidity was one of the most reliable predictors of fire danger in the forest.[11]

Hofmann's research had an immediate effect on forest practices in research and industry. Forest districts soon had tools that translated levels of humidity to levels of fire danger. Initially the Forest Service established a standard of 30 percent relative humidity as the danger point. One veteran firefighter later recalled, "Gradually we got the idea that perhaps 30 percent was not so much the danger point as the disaster point." The benchmark for relative humidity was soon raised (fig. 3.3).[12]

By linking relative humidity with fire potential, Hofmann's work gave foresters a predictive power over their enemy, fire. This was one of many steps throughout the century to predict, control, and limit fire in the national forests. The ecological role of fire in forest

*Fig. 3.3. The instrument
station established at Wind
River to study fire weather,
1929. (Courtesy of USDA
Forest Service.)*

development would not be studied in depth until the end of the 20th century, after decades of fire suppression and the effects of a changing climate created new problems in western forests.[13] But in the early 1900s, the biggest challenge was to control fire and replant burned and cutover forests.

Because the task of reforesting the burns was so vast, researchers at Wind River studied both planting and natural regeneration as ways to restore the region's forests. They needed to know what conditions limited the natural reforestation of cutover and burned land and what forest practices could be developed to better ensure a second crop of trees after harvest. Hofmann examined how Douglas fir naturally revegetates land laid bare after clearcutting or fire. In 1917 he created a transect of small plots across a swath of the Wind River valley. A year earlier the Wind River Lumber Company had removed trees from ridge to ridge with overhead cables. Down the slopes, across the valley, and up the other side, Hofmann recorded the establishment of every tree and seedling that appeared within the transect plots. The U-shaped transect made it possible to compare seedling growth on the northeast- and southwest-facing slopes, and at various distances from the edge of standing timber.[14]

Forest scientists at Wind River continued to measure the growth of every seedling within the transect for the next several decades. After 20 years, the cool, moist, northeast-facing slope had fully reestablished within a quarter mile of the forest edge, whereas the regenerating stands on the hot, dry, southwest-facing slope were sparse and irregularly spaced and located only within a couple of hundred feet of the forest edge. These findings led scientists to conclude that clearcut areas should be between 200 feet and a half mile wide, depending on the site, to ensure that seeds from adjacent standing trees would regenerate the area.[15] Time, however, would rearrange some of these findings. By the mid-20th century, growth had slowed on the crowded northeast-facing slope and heavy snow had broken some trees and toppled others. During this time, trees on the sparsely stocked southwest-facing slope had continued to grow, and by the early 21st century, they were larger than trees on the overstocked northeast-facing slope.[16]

The Wind River transect was one of many study sites that measured the regeneration of trees. As early as 1911, E. T. Allen had seen that the future of commercial forestry in the Northwest would be in the management of second growth.[17] Much of the early work at Wind River was meant to urge private timber owners to hold on to their land and replant their forests after cutting and to grow forests as a crop instead of mining them and moving on.

One method to boost production from young forests was borrowed from Europe and involved thinning out trees that had no commercial value to create more growing space for the remaining trees. Hofmann began an experiment to test this so-called precommercial thinning in 1919 near Martha Creek at Wind River. He established three plots where young Douglas fir had regenerated in a dense stand following a clearcut nine years earlier. He thinned two plots to 8-by-8-foot spacings. In one, he left only the trees that created an exact 8-by-8 spacing, even if he had to keep poorer trees to do so. In the other plot, he chose the most dominant tree within a cluster and thinned an 8-foot circle of trees around it. He left a third plot unthinned to use for comparison.

The experiment outlasted Hofmann and would defy easy conclusions in the future. In the 1930s silviculturist Walter Meyer added to the study three more plots that were thinned at age 22 and again 20 years later. At first, the plots thinned around a dominant tree gained the most timber volume, outgrowing both the unthinned plots and the plots thinned to exact spacing. But by the middle of the century, when the trees were about 40 years old, the results began to shift.[18] The plots with exact spacing began to show better growth, even though they had initially contained many less-vigorous trees. Over time these poorer trees died, leaving more room for the remaining trees. More experiments with spacing followed, eventually prompting the timber industry to adopt the widespread practice of precommercial thinning to enhance the greatest amount of growth in the shortest amount of time.

Thinning studies focused on places in the forest where natural regeneration had been successful and where a flush of young trees competed for space and light. But there were many places around Wind River where trees did not reestablish, where cutover forestland had become a tangle of treeless brush. On these lands, Wind River scientists planted nursery-raised seedlings. Although much of the work accomplished by nursery scientists at Wind River involved pioneering studies in seedbed design, transplanting, and seedling shipment, the forest scientists contributed to the choice and collection of seed that promised the best results.

As the nursery expanded its production to millions of seedlings for planting on thousands of acres, it faced the problem of needing trees that were suited to different conditions across the Douglas fir region. The nursery required a huge volume of seeds, but from what kind of trees and sites should seeds be collected? Wind River scientists needed information about how seeds grow, what seeds grow best at specific locations, and, ultimately, which seeds make the best trees in the long run. Such questions would take a lifetime to answer. Still in his twenties, Munger launched what he assumed would be a 40-year study of Douglas fir seed from various sources throughout the region.

The study is remarkable for many reasons, particularly for the details of heredity and genetics it demonstrated. The general principles of genetics had been described only a decade earlier when European scientists rediscovered records of the experiments by 19th century monk Gregor Mendel. Since then, few scientists had applied these new concepts to such long-lived organisms as trees. Munger noted, "Forest tree heredity is perhaps a rather uninviting field for the geneticist, unless he have [sic] uncommon patience, for the lifespan of a tree is so much longer than that of a man."[19]

Munger was uncommonly patient and his study was meticulous. In the fall of 1912, he sent out a crew to collect cones from 13 locations in the western Cascades of Washington and Oregon. He had selected sites that would provide a range of altitudes from 100 to 3,850 feet and instructed his crew to select cones in each location from trees young and old, large and small, with and without infection, in good and poor growing sites, and in dense and open stands, and to record all these details with each cone collected. His crew made sketches of each tree in each group to illustrate its form and cone placement and recorded the traits represented in each collection. To eliminate "the personal element and the chance of bias toward this or that conclusion," they recorded traits in code and assigned a number to each collection of cones. Through this strategy, Munger obtained seeds that could be traced to known trees with recorded, comparable traits.[20]

At the nursery, Munger subjected the cones to a battery of tests and immediately found differences among them (fig. 3.4). He found that large cones from open-growing trees bore larger seeds that were more likely to sprout, regardless of the tree's age or size. He noted that seedlings sprouted from seed collected at different altitudes sent

Fig. 3.4. Cones collected for the Douglas fir heredity study drying in the Wind River Nursery. (Courtesy of USDA Forest Service.)

out buds of new growth at different times. With this vast genealogy of Douglas fir seed, Munger wanted to sort out all the differences he could detect in various sources of seed and identify the most promising characteristics of seed trees that should be left for natural restocking. He envisioned a compendium of information about the source of Douglas fir seed for artificial reforestation and rules for cone collecting for the best possible nursery seedlings. Munger suggested the research possibilities to Hofmann, and soon after he arrived at Wind River, Hofmann and nursery researcher C. P. Willis took over the study.

Hofmann and Willis examined the reproduction of thousands of seeds from all the known sources. They found that seeds did best if collected from locations either as cold as or colder than where they were planted. They recommended avoiding seeds from high elevations or from low-quality sites. Surprisingly, they found that selecting trees for seed based on the tree's age and health or on soil type made no difference in seed production or seedling vigor. Hofmann later applied these findings to suggest harvest methods that would leave the forest with the best chances for natural regeneration. He concluded, "When selection cutting is done, thrifty well-formed trees will be left for future growth, and these will also become the most desirable seed trees."[21]

Munger's heredity experiment continued. From his geneology of seeds, Munger now had thousands of seedlings with known heritage ready to plant. He directed the crew to locate six planting locations

Fig. 3.5. The Wind River Douglas fir heredity study, established in 1915. (Courtesy of USDA Forest Service.)

throughout the region, including Wind River, plant representatives of each family in each location, and compare the results over time. Each tree in each plantation was marked with the code of its individual inheritance (fig. 3.5).

At this time and throughout much of the first half of the 20th century, plant and animal species were being introduced across the countryside with little or no concern for their adaptability or potential risk to native communities. Munger's focus on native trees and his care to catalogue their heritage and monitor their success was unusual for the time. He was testing the adaptability of particular races to various sites, with the intent of learning which races would succeed where, unlike so many other fish and wildlife introductions that proceeded without apparent testing.[22]

Although his study preceded the modern understanding of genetics and statistics, Munger's heredity plots continued to reap important— and changing—results throughout the century. Studies of heredity were relatively new at the time, but Munger made remarkably complete documentation of each tree's genetic lineage and designed his study with six replications, a strategy specifically meant to describe variation within each 5-acre plot at each location. Although the study lacked the randomness necessary for most modern statistical analyses, the long-term value of these plots cannot be overstated.

Ninety percent of the seedlings survived their first year after planting, but within a few years, differences among the genetic races began to show up. For example, Douglas fir saplings grown from seed collected near Darrington, in western Washington, outpaced most of the other seed families in all locations, demonstrating racial differences across the vast range of Douglas fir. The heredity study provided a mechanism for choosing and improving trees for planting in particular locations. By the 1940s new patterns began to emerge in the experimental plantations. High-elevation stock survived the rigors of high-elevation sites better than those from lower elevations. Eventually the coastal race died in every plantation except the coastal site, and Cascades seed did best at Cascades sites. The best regeneration came from using local seed; and the use of local seed sources dominated nursery practices for decades to come.

Although a minor component of westside forests, ponderosa pine was a significant timber tree in the intermountain West, and one that foresters wanted to grow better and faster. It was ponderosa pine that Munger had headed west to study, and in 1926 he contributed to a large-scale investigation of the tree's racial differences throughout its range. Munger tested 10 different families, or provenances, of ponderosa pine collected throughout Oregon and Washington and planted at sites at Wind River and elsewhere in conjunction with similar studies in Arizona, Idaho, and later New Zealand. The studies continued for more than 50 years and revealed again the long-term adaptability of local seed to local conditions.[23]

The enormous database that accumulated for all the trees in all the sites for both the Douglas fir and the ponderosa pine heredity studies represented the time and dedication of generations of scientists. Wind River forest assistant William Morris noted that it took three working days for one man to create the tags and more than a week to retag 2,600 trees in one study plot at Wind River.[24] Time continued to shape the heredity studies. Some genetic families grew much faster than others, only to be knocked back by fire, ice, or disease. A cold snap in 1955 killed almost a quarter of the largest Douglas fir trees on a coastal site near Hebo, Oregon, most of which were grown from high-elevation seed stock.[25] Toward the end of the century, Roy Silen, the station's first forest geneticist, was still collecting data from both these heredity studies.

While much emphasis in the early years was on regeneration and planting, the question still lingered about how forests contribute to water supplies. The concern first articulated by George Perkins Marsh was reflected in the 1897 Organic Act establishing the national forests, in part "for the purpose of securing favorable conditions of water flow."[26] Studies in 1916 examined the effects of melting snow at the Wind River and two other watersheds in the eastern Cascades. Researchers found that snow remained longer under the forest canopy than in open areas, and although there was less snow accumulation in the forest, they found that the snow in the forest lasted later in the season, slowing the thaw and reducing the threat of floods (fig. 3.6).[27] Later, Wind River researchers found that snow on clearcut land melted before snow in partially cut stands,[28] and snow lasted even longer in uncut old-growth

Fig. 3.6. Drifts of snow lingering on north-facing slopes of the Wind River site in A. A. Griffin and Charles Kraebel's study of the effects of melting snow, 1917. (Courtesy of USDA Forest Service.)

stands. Such studies recognized the connection between the condition of the canopy above and the snow on the ground. This connection would be examined again toward the end of the 20th century, when studies in the region documented the effect of rain on snow and how it can increase flooding in clearcut areas.[29]

Despite its recognition in statute, water was a less obvious forest product than wood. When the Great War broke out in Europe in 1914, the world needed wood. Nearly the entire trench system of the Western Front was built from wood. The Allies needed millions of board feet for bridges, barracks, ammunition boxes, rifles, and coffins. The country's limited shipping capacity made export impossible, so instead, the United States exported foresters. More than 30,000 American foresters, millworkers, and woodsmen served in Europe.[30] Several Wind River research assistants joined the 20th Engineer Regiment and sailed to France to help establish lumber mills and provide timber for the Western Front.

Hofmann stayed at Wind River and kept the research studies going, sometimes single-handedly, at the experiment station. Despite hardships during the war and occasional clashes with the more conservative Munger, Hofmann was a prolific researcher who was widely published in scientific journals. "The Natural Regeneration of Douglas Fir in the Pacific Northwest," Hofmann's 1924 publication, was one of a handful of studies that pushed the knowledge of American silviculture to professional levels, representing years of investigation and "scientific achievement of the first order," according to historian Andrew Rogers.[31] But one of the signature hypotheses in Hofmann's understanding of Douglas fir regeneration would eventually prove to be wrong.

An important aspect of Douglas fir harvest and regeneration was the fate of seed that falls to the forest floor. Allen claimed that Douglas fir seed could fly on the wind for up to a mile before falling to the ground; therefore, seed trees could be left up to a mile apart.[32] Munger offered a more conservative estimate. He estimated that Douglas fir seed could fly twice the height of the tree, so seed trees needed to be within about 350 feet of a clearcut area to reestablish the stand.[33]

Hofmann considered the question from a different perspective. He began with the assumption that "distribution of seed by wind is limited . . . to a distance of five chains from green timber . . . usually only a few chains during each seeding generation."[34] A chain equals 66 feet, so even at 5 chains, Hofmann assumed that seed flight was limited to 330 feet, which was consistent with Munger's estimate. And yet Hofmann observed regeneration of Douglas fir for a mile or more from any visible seed source. Therefore, Hofmann conjectured, for a tree to grow, seed had to be left behind, buried in the duff of the forest floor after the trees were harvested. Hofmann examined forest duff and found what appeared to be dormant seeds, and he noted that where duff had burned, no seedlings grew. Hofmann also examined forest stands that had grown up following wildfires and found trees that had established several years after the fire, with no seed trees left standing within 300 feet.

"The proportion of age classes tells the history of the reproduction," Hofmann wrote. He found that only a small number of the total regenerated Douglas fir seedlings had sprouted the first year after a fire. Far more had germinated between two and six years after the fire, with germination apparently dropping off after six years.[35] So Hofmann assumed that the Douglas fir seedlings he had found farther than 300 feet from seed trees would have sprouted from seed stored in the duff. He set out to test his hypothesis by storing seeds of many species, including Douglas fir, in test plots on the forest floor under the forest canopy. After just one season, he recorded that no germination had occurred, which he assumed showed favorable storage conditions. He concluded that the seeds of all the tree species were still viable.[36]

His conclusions had particular importance for the timber industry and regeneration efforts. According to Hofmann, when the forest was removed by fire or cutting, stored seed would be left behind in the duff, awaiting favorable conditions in which to germinate. If seed were stored in the forest floor, there would be no reason to leave trees standing to provide seed for reforestation, no reason to leave anything standing that could burn. Richard Rajala suggests that Hofmann's theory strengthened the argument for boundless clearcutting in the Douglas fir region.[37]

Unfortunately, the theory would be proved wrong.

Chapter Four
A Forest for the Long Term

⚜

The most progressive nations are without exception those which have engaged most extensively in research.

—Earle Clapp[1]

Research at Wind River languished following World War I. By 1921, after a series of resignations and budget cuts, Julius Hofmann once again found himself the only professional forester at Wind River, attempting to keep experiments going single-handedly. During this time, Hofmann was finishing his Ph.D. at the University of Minnesota, and would occasionally spend time there. It was during one of his trips to Minnesota that Hofmann met Leo Isaac, a young undergraduate forestry student. Isaac, who would eventually make his career at Wind River, remembered Hofmann as "one of these dashing, plunging sort of fellows . . . Right from the beginning, he insisted that I should come out and go into forest research. He arranged to have me come out as his assistant at Wind River. But a week or two before I got to come out here I got a telegram from him that his appropriation had been taken away and that he wouldn't have an assistant."[2]

Conflicts appeared to be mounting between the dashing, plunging Hofmann and T. T. Munger, the reserved, punctilious New Englander. Isaac remembered that "Hofmann got into some battles with Munger and some of the other foresters and they punished him by pinching his spare cash off."[3] Clashes between Hofmann and Munger would intensify with time, but before leaving the Douglas fir region, Hofmann had a role in crafting an important piece of forest legislation.

In the 1920s free enterprise and self-regulation began to replace the government-led reform of the earlier Progressives. Forest industrialists banded together in trade associations that gave them greater political influence to leverage government partnerships. Some, like former Forest Service chief William Greeley and former District Six forester E. T. Allen, made the jump from federal service to private enterprise and now advocated for strong partnerships between industry and government research.[4] In particular, they lobbied for a national forest policy that would increase federal fire protection without incurring federal regulations on industrial timber harvest.[5] Greeley courted the powerful support of Oregon senator Charles McNary and helped organize a trip for the legislative committee to visit the forests of western Washington and see the effect of fires on the federal forests. Hofmann was called on to help devise a forest policy that could be supported by legislation.[6] In 1924 the policy was passed in the form of the Clarke-McNary Act, which funded research in forest economics. The act also provided federal support for planting and fire protection without any regulation of harvest practices on private land.

A truce between Hofmann and Munger had been called while Hofmann worked on the Clarke-McNary legislation, but tension between them resumed immediately after the legislation passed. Although the research position with Hofmann had been cut, Isaac decided to go west anyway. Hofmann helped him land a job at the Chelan National Forest in northern Washington, and "at the first opportunity for expanded research they brought me down as a field officer there at Wind River,"[7] where the young field officer witnessed conflicts between Munger and Hofmann escalating over petty expenses.

Isaac recalled that if Hofmann wanted to do something, "he'd go ahead and do it whether he had money enough in his appropriations or not." That included paying for the construction of the government-owned house Hofmann had built at Wind River. "I was between two fires because there was very bitter feeling between Munger and Hofmann," Isaac said.[8] The exacting Munger finally called Hofmann to the carpet for overbilling a small travel expense. Hofmann refused to make up the difference. The issue—half the cost of a hotel room—was trifling, but it became a showdown.

Isaac recalled, "He [Hofmann] claimed he had made up the difference on the meals. But they told him to correct it and he refused to do it, he was that stubborn."[9] Hofmann was given a six-month disciplinary furlough; he immediately resigned.

With his hands full as director of silvics for District Six, Munger needed someone to fill in for Hofmann and oversee research at Wind River. He called Isaac in "to tie up loose ends at Wind River" and offered the young researcher Hofmann's house. Hofmann had paid for part of the construction of the house and seemed to be in no hurry to depart. With no other options, Isaac and his family squeezed in with the mercurial Hofmann and his family, sharing quarters for six long weeks. "We told him not to hurry, but we didn't know it was going to last that long," Isaac later remembered with good humor.[10]

In his 10 years at Wind River, Hofmann had done pioneering work on regeneration and seed characteristics, fire studies, and genetics. He was a prolific researcher, publishing more than 30 articles in the popular press and scientific literature during his years at Wind River. Hofmann included a list of cited references at the end of his professional papers; he was the first Wind River scientist to credit previous work by fellow researchers. The Ph.D. he earned at Minnesota was the first ever awarded from an American school of forestry. But much of Hofmann's accomplishments at Wind River would be overshadowed by his hypothesis of seeds stored in the duff, which Isaac would soon debunk.*

* After Julius Hofmann quit the Forest Service, he went on to be assistant director of the Pennsylvania State Forest School at Mont Alto. This was the first public forestry school in the United States. From there he moved south, taking with him a full complement of students, graduate students, and professors from the Mont Alto school, to open a new school of forestry at North Carolina State University. His entrepreneurial ways continued with the unorthodox establishment of a teaching forest in North Carolina, procured through a complicated choreography of tax lot revisions. His program at NCSU grew to include several more teaching forests all around the state, the largest, and the first, came to be known as the Hofmann Forest.[11]

Hofmann had barely packed his bags before federal funds came through to increase research funding in District Six and to establish the Pacific Northwest Forest and Range Experiment Station for forest research throughout Oregon and Washington.* Munger was named its director in 1924, with a staff that was spare but effective. Richard McArdle, who would eventually rise through the ranks to become chief of the Forest Service, was appointed junior forester from the Civil Service eligibility list. Among the four forest assistants at the new station was Bob Marshall. Marshall would later become an eloquent and influential advocate for wilderness areas, but at Wind River, Isaac viewed him as "a very odd chap." According to Isaac, Marshall broke practically every instrument and tool Isaac gave him to work with. Isaac recalled that when influential visitors came up to visit Wind River "he spent most of his time running backwards in front of them snapping their pictures, and picking stuff up out of their hands and carrying it for them, and that sort of thing . . . He was very queer, very odd."[12]

As director of the new station, Munger tailored his approach to fit the politics of the time. Although his interest was in long-term forest research, his target audience was clearly the timber industry of the West. "This forest experiment station is created to serve the industry for the public welfare," he said as he outlined his objectives for the new experiment station before the Pacific Logging conference in 1924. As Munger described it, part of that service was to improve on nature. Research, he asserted, could make Douglas fir forests "yield more for their age than the virgin woods can." Munger's dedication to the ideals of scientific forestry was audible in his language. "There is little satisfaction in working with a decadent old forest that is past redemption and calls for nothing but removal," he told his congregation.[13] In almost biblical images, he described a research agenda for the new

---

* The Pacific Northwest Forest and Range Experiment Station was renamed the Pacific Northwest Research Station in the 1990s. In this book we refer to it as the PNW Research Station. We refer to scientists affiliated with the station as PNW researchers.

experiment station that would help remove profligacy and decay from the northwest forests, where a new, more productive forest would rise from the ashes of the old.*

Yet Munger was not a simple mouthpiece for the agency's embrace of industry concerns. "Man's industry in converting the forest into useful products is having a cataclysmic effect on land productivity," he wrote in 1930.[15] Since the days of the fur trappers, industry had been moving into the Northwest with the notion that resources in America were inexhaustible. Munger saw that they were not. He had arrived in the Pacific Northwest when timber speculators were taking the best land, small mills were illegally cutting the best trees, and fires were periodically charring the landscape. Munger believed that the forest was in need of salvation; it had to be saved from the greed of timbermen and from its own wasteful native ways. He also believed that science could remake the forest into a production system that was thrifty, economical, and renewable. Munger said, "Timber has been regarded as an unrenewable resource, like coal, instead of as a crop which by proper management could be harvested periodically."[16]

With a brand-new forest experiment station in the Douglas fir region, Munger's own corps of forest scientists set out to reduce waste and to redesign the forest to grow a predictable crop of trees for timber. They adapted agricultural practices to encourage forest managers to change from mining native timber to regenerating a sustainable, rotational crop of trees. Future generations would criticize those agricultural practices as demanding an unnatural uniformity from complex forest systems. But agricultural practices that promised a sustainable crop were a rational response to a cut-and-run timber industry that left broken communities and spoiled landscapes in its wake. Within the context of the times, this agriculture approach to forestry suggested an ethic of good husbandry and long-term care for the land.

---

* Munger often colored his speeches with anthropomorphic images. In a speech dedicating the Peavy Arboretum at Oregon State College and honoring Dean Peavy of the College of Forestry in 1926, Munger called the trees in the arboretum "the offspring of his hard work . . . every tree should be taught to call [Peavy] affectionately 'Daddy.'"[14]

In truckloads of boxes and files, the administration of research in District Six moved from Wind River to Portland in 1924.* The new PNW Research Station would continue experiments underway at Wind River, which was designated as a fieldwork center "for fire studies and reforestation projects" with Isaac as its director.[17] Isaac could not have been more different than the man he was replacing, nor less like the man who would be his new boss. Whereas Hofmann had been arrogant enough to make up his own rules, Isaac was exacting in his observations and reporting. And compared to the cool, stern Munger, Isaac was fun-loving and friendly with colleagues. Roy Silen, a forest geneticist whose 50-year career began at Wind River, later remembered his old bosses as distinctly different personalities: "Munger was a ramrod of a fellow. He was a good scientist . . . He was usually curt. He'd answer in very short sentences . . . I think Munger felt he was always a cut above." In contrast, Silen remembered Isaac with great fondness. "Isaac was the fatherly type . . . He was really the kindest person that we had around the station." Silen also remembered Isaac as an extraordinary scientist and teacher: "He would try to make sure you could understand the finer details. We would go out and work all day in the field, and all the time there was a conversation about something or other. He had a sense of humor. I think he was among the better examples of a person who could see through the character of other people. He was special, but somehow, he wasn't appreciated by the people at higher levels in the station, the regional office, or the Washington D.C. office."[18]

One of the first studies Isaac pursued was to test Hofmann's assertion that new forests could spring from seeds stored for years in the leaves and litter on the forest floor. Although well-established in the literature, thanks in part to Hofmann's own prolific writing on the subject, what had long bothered Munger was the observation that vast areas were not reseeding. Land would stay bare for years after a burn, until there was a good crop of seed in the distant stands. He assigned Isaac the task of finding out why.

* After 1930 Forest Service districts were renamed regions, with subunits of forests and ranger districts.

Although Hofmann had tested buried seeds and found that they did not sprout after a year, he reported in his 1917 publication that the unsprouted seeds were still viable and would eventually sprout when the conditions were right. He ended his experiment at that point, leaving forest managers with the reassurance that they could rely on revegetation from seeds buried in the ground. It was not until 1924 that Hofmann reported that rodents had invaded the test plots and may have eaten some of the seeds.[19]

Isaac revisited Hofmann's experiment in a study that would last nearly a decade and take him down to the forest floor and up to the sky. At Wind River, he buried Douglas fir seed in rodent-proof containers at three different depths in the duff, and tested seed viability every year for three years. Finding little germination the second year and none the third year, Isaac repeated the experiment (fig. 4.1).[20] Isaac said, "I repeated the germination tests three times before I published anything on it, to be doubly sure, because I didn't want to hurt Hofmann, for one thing. And I didn't want to destroy an established theory if it had any foundation of fact in it."[21] Isaac's careful study concluded that old

Fig. 4.1. Leo Isaac checking for sprouting seeds in rodent-proof containers during his reexamination of Hofmann's theory of seed storage in the duff, 1926. (Courtesy of USDA Forest Service.)

seed stored in the duff was not a source for regenerating forests. But if seedlings did not sprout from buried seed, then what was the source of scattered trees sprouting on burned lands?

In 1926, as Charles Lindbergh was about to capture the world's imagination with airplane flight, Isaac launched his own test flights. Isaac observed that the winglike blade of the Douglas fir seed was similar to an airplane propeller, and noticed that the seed would spin through the air, lifted aloft and forward before falling to the ground. He needed to know just how far those propellers could carry seed in the wind. Hofmann, like Munger, had estimated a maximum seed flight of just over 300 feet, but neither scientist had tested that widely accepted assumption.

Isaac designed an ingenious test using a big silk kite, a small oatmeal box, and an open field of snow. He filled the oatmeal box with a known number of seeds and marked off a huge grid across the snow-covered field. With the kite, he lifted the Douglas fir seeds aloft to a specific height and released them to the winds to fly and settle across the field. He recovered each seed and recorded its landing spot on the grid. Isaac tested different heights of release, and on days with little breeze, he used a balloon to carry the box of seeds (fig. 4.2).

Isaac found that the higher the release—or the taller the trees—the farther the seed would fly; doubling the height of release more than

Fig. 4.2. Leo Isaac used a kite and an oatmeal box to launch Douglas fir seeds to measure how far seed is blown by wind, 1926. (Courtesy of USDA Forest Service.)

doubled the distance of seed flight. Seeds from a good crop could be expected to travel at least 900 feet from the edge of the forest—three times the distance Hofmann and Munger had each assumed.[22] Isaac's findings began to explain limits to natural regeneration. Seeds could fly, but their reach was limited. Many burned and cutover lands were so extensive that there were no trees anywhere close enough to provide a source of regenerating seed. Even where seed trees were available, regeneration was spotty; up to 95 percent of the seed crop never germinated.

Working with McArdle, Isaac conducted a sweeping study of the ecological aspects of natural regeneration using Wind River studies and observations throughout the Douglas fir region. They found that seeds were more likely to sprout on bare mineral soil, but not on dark, charred soil. Although burning leftover material was necessary to open mineral seedbeds, it could leave the soil too black and too hot for germination. And leaving too much unburned slash invited subsequent fires that could wipe out a regenerating stand of trees. In either case, the land required years of intensive fire protection while the second forest was being established. Considering all these factors, Isaac and McArdle recommended leaving strips of uncut timber one-quarter to one-half mile wide to facilitate natural regeneration within the harvested areas.[23]

The concept of growing trees and calculating rates of return on investment met with skepticism in the private sector. Private forestland owners needed documentation if they were to be convinced that trees could be grown profitably as a continuous crop. Munger had begun that documentation when he established permanent plots in stands of young Douglas fir in 1910. According to Munger, permanent sample plots would provide "a record of the life history of the stand and of the actual growth that was taking place."[24] With meticulous measurements of growth over time, he expected to learn "what nature would do if undisturbed by fire, insects, fungi, and other enemies."[25] Munger intended to provide long-term information for the taxpaying landowner to predict the growth and yield of Douglas fir. Data would be compiled into yield tables that a forest landowner could use to predict the amount of timber he could expect from his second-growth stands and calculate the best time to harvest trees. "The business

Fig. 4.3. Researcher collecting growth and yield data on a sample plot, using an increment corer to determine tree age and calipers to measure tree diameter at breast height, 1925. The information collected was reported in Bulletin 201. (Courtesy of USDA Forest Service.)

of growing successive crops of timber . . . can be figured out in a business-like way by anyone who has faith in the future and is willing to prophesy the size of timber that will be in demand and the stumpage it will bring."[26] So Munger, the scientist and preacher's son, advocated forest management based on numbers, faith, and prophesy.

Such yield tables had been developed in Europe and were used extensively in thinned, multi-aged forests, but forest managers needed new tables to calculate potential growth in young stands of single-aged, unthinned Douglas fir. McArdle took up the study with Walter Meyer, another Yale graduate who joined the station in 1925 as an assistant silviculturist. McArdle and Meyer expanded the data collection that Munger had begun at Wind River to include over 2,000 plots at 261 locations throughout the Douglas fir region (fig. 4.3). Now forest managers could decide the best age at which to cut timber and calculate the sort of profit they could expect to make.[27]

Robert Curtis, a PNW research mensurationist, reflected years later that "this was a mind-boggling achievement for a time when travel was slow and when data summarization and analysis depended on the slide rule, the mechanical adding machine, and graph paper."[28] McArdle and Meyer published their growth and yield tables in 1930 as U.S. Forest Service Technical Bulletin 201, "The Yield of Douglas Fir in the Pacific Northwest." The publication, referred to as Bulletin 201, had immediate application to forestland owners throughout the region, offering quantitative evidence of the enormous productivity of Douglas fir.[29] The numbers were convincing: money could be made from young-growth trees. Timber management on private land began to change from "cut and run" to "stay and invest," although this pattern would switch again in the 21st century.

This new understanding about the potential productivity of young, natural stands prompted new interest in ways to coax more production from young-growth stands of Douglas fir. Experiments with optimum spacing had been explored by researchers at Wind River, beginning with Hofmann and his thinning studies at Martha Creek. The research question was a practical one: what effect did the spacing between growing trees have on their height, diameter, and yield of timber? To test this question, in 1925 Isaac chose a site at Wind River that had been burned and then reburned, clearing the site of all but brush and weeds. He planted blocks of young Douglas fir seedlings from the Wind River Nursery at spacings from 4 feet to 12 feet apart and measured their progress every few years. In the first five years, the narrowly spaced trees grew taller than their wider-spaced neighbors. But in the next five years, the more widely spaced trees began to outpace the initial spurt of the more crowded plantings. Continuing years took their toll, as disease, snow, and ice whittled away at the spacing study plantings. This relationship of spacing to growth would be revisited many times over the century, as would the question of thinning forest stands to increase timber volume.[30]

It is important to note that the ideal forest that these scientists wished to re-create was modeled after the forests they found growing naturally in many places throughout the Douglas fir region. Nature had sculpted extensive stands of even-aged Douglas fir from the natural force of fire. In their earliest reports, both Allen and Munger had observed Douglas fir's affinity for fire, the thick fireproof bark on the oldest trees, and the regeneration of seedlings on burned, bare soil. But in the 1920s, wildfire was considered more of a problem than a natural process in the forest, and studies of fire focused primarily on its prediction and prevention. In his 1924 statement of purpose for the new PNW Research Station, Munger estimated that "fire control is 75 to 90 percent of forestry" and that forest research would discover "how to control the forest fire plague."[31] Certainly fire prevention was on the mind of the regional forester, who supplied the Wind River researchers with three meteorological stations to further their studies of the relation between weather conditions and fire danger.

Hofmann had established the correlation between low relative humidity and high fire risk. Gael Simson, who had come to work at Wind River during Hofmann's last year, continued Hofmann's studies of fire and weather. During World War I, Simson had worked with the naval radio laboratories, and he brought his experience with radios and static electricity to the calculation of fire risk in the forest. At Wind River, he examined the relationship between static electricity and relative humidity and used those correlations to refine fire prediction. Years later, Munger remembered Simson, "who was not a forester, but good deal of a genius."[32] Part of Simson's genius was using static on the radio to predict the approach of lightning storms, a major cause of forest fires throughout the West. To detect slight changes in relative humidity, Simson built a static electricity station at Wind River, which was basically a radio receiver meant to capture the static rather than the broadcast. With two 115-foot towers and a 10-foot square loop aerial, Simson was able to measure changes in static electricity and provide an index of relative humidity. This proved to be a reliable way to predict the fire danger in the forest each day. Simson's work led to the development of small, battery-powered portable radios that lumbermen and researchers alike could carry with them in the woods. The devices proved to be doubly useful as fire predictors and as portable communication tools, the forerunners of the walkie-talkie.[33]

Still, fire continued to be a threat and a puzzle to researchers in the Douglas fir region. Even fast-growing healthy forests were susceptible to this natural disturbance. The head of Forest Service research questioned whether thoroughly effective fire protection could ever be obtained without more fundamental fire research. McArdle quoted him in 1927: "The lack of satisfactory progress goes back, among other things, to a lack of fundamental knowledge."[34]

Fire history throughout the region had shown that the forests of western Washington and Oregon burned most frequently and with most intensity in the late summer and early fall, at the end of the long dry season that characterized the Douglas fir region.[35] To understand why, Wind River field assistant William Morris measured the moisture content of everything on the forest floor—logs, snags, shrubs, herbs, and needles—at Wind River several times weekly throughout the summers of 1935 and 1936. He found a gradual decrease in moisture into September, which helped to explain why fires burn hotter later in the season. He also measured changes in moisture in all the flammable material throughout the day under different forest cover conditions. Not surprisingly, he found that the more cover above, the less moisture loss below.[36]

In just the way Simson had developed a tool to use in the woods to predict coming fire weather, Morris wanted to develop a tool to help foresters measure the flammability of the forest by measuring the moisture level in the fuel. He began with simple sticks. With help from assistant forester Donald Mathews, Morris tested many shapes and kinds of wood that would represent the different fuels that lay on the forest floor, ready to carry fire. They found that Douglas fir sticks lost moisture relatively slowly and represented heavier fuels in the forest. Smaller sticks of ponderosa pine lost moisture relatively quickly and represented the rapid changes that can take place in fine forest fuels from hour to hour. The fire sticks they developed used both Douglas fir and ponderosa pine to indicate moisture level, and therefore flammability, of fuels on the forest floor. Moisture content, determined simply by weighing the sticks, provided a measure of fire danger in the forest, a method still widely used throughout the West.[37] Today's Fire Danger Rating sign, the ubiquitous dial that indicates the day's fire danger to passing motorists, uses measures of relative humidity and fuel moisture, both of which were pioneered by researchers at Wind River.

Clearly the forest was flammable, but what was the source of the flame? In addition to lightning strikes and runaway slash burns, fires at the turn of the 20th century had been ignited by sparks flown from wood-burning locomotives or land-clearing homesteaders. By the mid-1920s, people increasingly traveled through the woods in automobiles, and the popular image of the urbane traveler always included a cigarette. "Even the ubiquitous cigarette butt comes under cold scrutiny of the scientist in the course of his persistent efforts to aid the fire fighter in keeping fire out of the woods," Simson wrote in 1926. "To find out relative frequency of fires caused by cigar or cigarette butts . . . will prove of material aid in various ways, for example in drafting smoking regulations within the forests."[38]

In these years before Smokey the Bear, Mathews studied how smoldering cigarettes and matches affected forest fuels at known moisture levels. He found that the combination of smoldering cigarettes on rotten wood posed the greatest risk of forest fire. "Cigarettes and rotten wood make the most dangerous combination . . . [therefore] it is desirable to eliminate all powdery, punky, rotten wood from trails, roadsides, campgrounds, and other parts of the forest where smokers might possibly discard a burning cigarette."[39] The result would be a park-like forest swept clean of the woody debris that aided and abetted the enemy, fire.

Snags, the skeletons of dead trees left over from fires or killed by disease, were another symbol of waste and danger to the first Wind River researchers. Although later generations of forest scientists would document the value of snags, particularly to forest wildlife, Munger had only contempt for them when he wrote this description: "They stand, fringing the skyline like the teeth of a broken comb, in mute defiance of wind and decay, the dregs of the former forest, useless to civilization and a menace to life of man and of forests . . . Snags deserve outlawry, yet they continue to practice murder and incendiarism on millions of acres of fertile Douglas fir lands . . . The day will come when snags are banished altogether from Douglas fir logged-off land."[40]

In an effort to banish these snaggle-toothed enemies, researchers tested a variety of ways to burn or blast snags in the Wind River valley. They found that loading dynamite into the side of a snag could explode the offensive structure most effectively. Many years later, forest scientists would document the value of both standing deadwood and downed woody debris to the forest ecosystem, but throughout the

Fig. 4.4. Snags, considered a menace in the early 20th century, exploded with dynamite, date unknown. The Wind River Nursery and buildings are in the background. (Courtesy of USDA Forest Service.)

mid-20th century, downed and rotting wood was thought to be a waste and a hazard, something to be eliminated from the regulated forest (fig. 4.4).

Despite efforts to reduce Northwest forests to a park of tidy, fast-growing timber, Munger realized the importance of understanding the dynamics of the natural, "virgin" forest. An article by ecologist V. E. Shelford advocating virgin timber reservations set Munger thinking about the stands of old-growth Douglas fir that even in the 1920s seemed to be disappearing fast. Munger proposed a 280-acre stand in Wind River as a study site and a virgin timber reservation. The regional director of forest management responded, "We've got twenty million acres of virgin timber in the National Forests; why set up this special area?" The supervisor of the Columbia National Forest opposed the idea as well, saying the tract contained cedar he needed for shakes on the Wind River buildings. Undeterred, Munger sent his proposal to Washington, D.C., where the chief of forest management said, "Was it the policy of the Forest Service to withdraw some old timber from cutting just to let it rot?" Earle Clapp, the chief of research, countered, "The vision-less and reactionary memo from the Chief of Forest Management is noted," and he proceeded to put Munger's recommendation through.[41]

In 1926 the Wind River tract of old-growth forest became part of a fledgling program of research natural areas established as examples of forest types left in their natural condition "forever." Did they think that "forever" meant that the forests would not change? The value of studying old forests to better understand new forests complicated the agricultural approach to forestry. A 1929 book by I. W. Bailey and H. A. Spoehr outlined the role of research in reforming forest policy and challenged the use of the agricultural model as a basis of forest management. "Nature's own husbandry, as found in the remaining primeval forests, will ever be man's best teacher," the authors concluded. They described silviculture as "a more plastic and adaptable art than is agriculture" because silviculture focused on natural plant communities, rather than highly artificial and stereotyped units. The authors emphasized the need for long-term studies of forests, because "the time factor in forest experimentation is so long and the phenomena to be investigated are so complex and so extraordinarily variable." They concluded "that a clearly visualized and adequately financed program of basic scientific research should be initiated as soon as possible" to manage the earth's timberlands.[42]

At about the same time, a special committee of the Society of American Foresters, headed by Clapp, was examining the status of forest research. Good forest practices, they concluded, required research. The committee's report became the basis of new legislation authorizing a sweeping Forest Service research program. Clapp's leadership encouraged support from the forest industry for broad-based research. Again, Senator McNary was called on to help sponsor a bill, this time to support a broad research agenda, including a nationwide forest survey, and to expand the system of forest experiment stations. Congress passed the McSweeney-McNary Act in 1928, with its full

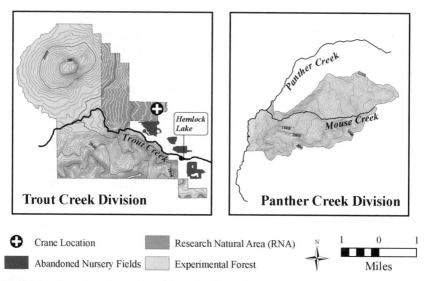

Trout Creek Division   **Panther Creek Division**

✚ Crane Location   ▆ Research Natural Area (RNA)

▆ Abandoned Nursery Fields   ▢ Experimental Forest

Miles

Fig. 4.5. Trout Creek and Panther Creek divisions of the Wind River Experimental Forest were established in 1932. (Map by Theresa Valentine, USDA Forest Service, PNW Research Station.)

appropriation of $3 million. According to forest historian Harold Steen, the law was so broad-based that the Forest Service would not need to seek additional legislation for research for another 50 years.[43]

The new legislation made it possible to complete Raphael Zon's vision of a network of experimental lands that would represent the dominant forest and range types throughout the Forest Service. Encouraged by Clapp and the new emphasis on research, Munger convened a committee to consider the establishment of two experimental forests, one for Douglas fir and the other for ponderosa pine. Pringle Falls Experimental Forest was designated for ponderosa pine in 1931. The next year, Munger assigned Isaac the task of locating the most suitable area for an experimental forest for Douglas fir. Isaac's search quickly focused on the forests around Wind River. He identified areas that would represent a broad range of conditions found in the Douglas fir region. Within a few months, Isaac proposed a 10,310-acre experimental forest in two divisions: The Trout Creek Division included a 6,500-acre parcel with areas burned by the Yacolt fire, unburned old forest up to 450 years old, and cutover lands, as well as the research natural area that had been established a few years earlier. The Panther Creek Division almost entirely comprised forests burned in the 1850s, with trees that were at this point about 75 years old. Isaac saw research value in these young, even-aged stands, which he said were "a type in which much private cutting is now being done and a size at which Douglas fir will probably be cut under management." In addition to these two divisions, Isaac included the 160-acre Martha Creek Flat, where some of the earliest experiments had been established by Munger and Hofmann.[44] At the same time, the Wind River Research Natural Area was expanded to just over 1,000 acres (fig. 4.5).

More experimental forests were established in the following years: Cascade Head in 1934 for study of coastal Douglas fir and Sitka spruce/western hemlock forests; Starkey Experimental Range in 1940 for range studies; and H. J. Andrews in 1948 primarily for watershed studies. In addition, the Port Orford Cedar and Blue Mountains experimental forests were established but later abandoned. With federal recognition of the importance of silviculture research and a 20-year track record of silvicultural research in the Douglas fir region, the Wind River Experimental Forest was officially dedicated in 1934. However, events far from the forests of western Washington would alter the direction of forest research. As the Great Depression settled over the nation, the country turned to its forests for material to rebuild the national economy and as a source of much-needed jobs.

Chapter Five
A New Deal for Forest Science

✁

It is clear that economic foresight and immediate employment march hand in hand in the call for reforestation.

—Franklin Roosevelt[1]

When Franklin Roosevelt accepted the Democratic nomination for president in 1932, the nation had been suffering bank failures and bread lines for three years. Lumber prices in the northwest were at historic lows. However, as historian Harold Steen pointed out, during the Great Depression of the 1930s, Forest Service researchers had more to do, and more money and manpower to do it, than ever before.[2]

Part of Roosevelt's New Deal was the Emergency Conservation Work Act, which called for the Civilian Conservation Corps (CCC), a million-man army dedicated to conservation. The Department of Labor selected enrollees; the Department of War operated the camps; and the Agriculture and Interior departments managed the work projects. Over the course of nine years, the CCC put three million young men to work in American parks and forests. Many worked for the Forest Service, planting trees, reducing erosion, and protecting forests from fire.[3]

Company 944, organized at Fort Lewis, Washington, was one of the earliest CCC groups deployed in Washington. In May 1933 an advance guard of the company arrived at Wind River to begin constructing a camp on Trout Creek, across from the Wind River Experimental Forest buildings.[4] It was the site of the former Wind River Lumber Company, which had closed in 1925; the land had been purchased by the Forest Service in 1929.[5] Under the CCC, it was known as Camp Hemlock and became home to as many as 170 young men, ages 18 to 25, and a staff of technical personnel, cooks, and the company officers (fig. 5.1).

The CCC not only brought manpower to the woods, but its building projects also created a new demand for lumber. Living initially in army pyramid tents, the CCC "boys" built Camp Hemlock according to the Army's standard specifications: rectangular frame buildings constructed from dimensional lumber. During their first season, they built four barracks, a mess hall, an infirmary, an education hall, officers' quarters and garage, and several shops and utility sheds, and they installed wood-burning stoves to heat the buildings. Over the next few years, they built a new office for the Wind River Experimental Forest and a regional training center that would be a meeting place for Forest Service managers and scientists for decades to come (fig. 5.2).

As the Wind River community grew, the existing power plant could no longer handle the additional demand. In 1935 the CCC began construction of a new powerhouse that harnessed power from a new dam that arched across Trout Creek. The Vancouver newspaper called this "the largest CCC construction project undertaken in the United

Fig. 5.1. Civilian Conservation Corps (CCC) at Camp Hemlock awakens at reveille, 1933. (The larger building in the background no longer stands.) (Courtesy of USDA Forest Service.)

Fig. 5.2. The Regional Training Center at Wind River, built by the CCC and completed in 1937, date unknown. (Courtesy of USDA Forest Service.)

States."[6] The powerhouse and its 183-foot dam provided electricity to over 30 Forest Service buildings at Wind River.[7] The project also created Hemlock Lake, a reservoir where the CCC developed a picnic area and boat launch, which quickly became a popular spot for residents of the Wind River valley.

Despite these ambitious building projects, the foremost goal of the CCC was the conservation of natural resources. At Wind River, the CCC worked in the nursery fields and constructed trails on Trout Creek Hill (fig. 5.3). The members built paths through the arboretum and placed water pipes there for fire protection. When a huge old Douglas fir blew down near Mineral, Washington, the CCC built a display at the arboretum entrance featuring a cross section of the fallen giant. The display allowed visitors to count the rings across the 15-foot diameter cross section, which showed that the tree had been almost 1,000 years old when it fell (fig. 5.4).

Firefighting got a boost from CCC manpower and the funds allocated by the Clarke-McNary Act. Throughout the area of the Yacolt Burn, fires had continued to flare up on old burned sites. Now with money and manpower, forest managers could implement firefighting strategies in the region. CCC crews felled snags, cleared brush, dug waterholes, and built roads and bridges that opened up more of the inaccessible parts of the rugged Cascade Mountains for fighting fires.[8]

Fig. 5.3. The CCC provided manpower for several projects at Wind River, including weed pulling at the Wind River Nursery, 1933. (Courtesy of USDA Forest Service.)

Throughout the Great Depression, the CCC provided much-needed laborers for research at Wind River. CCC recruits were paid pennies to prune branches to different heights on trees in a study to determine the effect of pruning to create clear-grained wood (fig. 5.5). Follow-up studies found that pruning up to one-quarter of the live limbs could produce clear logs with no loss in growth.[9] However, without the cheap labor of the CCC, pruning would prove too costly, and the practice was dropped after the CCC disbanded. It was not until the end of the 20th century, when clear, old-growth lumber was no longer available, that pruning limbs was again considered an economical management practice.

In parts of the United States beyond the Pacific Northwest, the 1930s brought devastating droughts and dust storms. The experience prompted new scientific thinking. Some scientists of the American grasslands were advancing the idea that changes in nature were aimed at a final climax condition, something like the ultimate and inevitable form of a fully developed organism. Their model of nature as an organism contrasted with another popular model, that of nature as a machine. The mechanistic model focused on measures of productivity and efficiency that could provide science with a blueprint to design a better world.[10]

Fig. 5.4. The Mineral Tree was nearly 1,000 years old before it toppled in 1930. A cross-sectional piece of the fallen Douglas fir is displayed near the Wind River Arboretum. (Courtesy National Archives and USDA Forest Service.)

Fig. 5.5. CCC men helped with forest research, including a study of pruning trees. (Courtesy of USDA Forest Service.)

The two models suggested different directions for forest science. The nature-as-organism model focused on a natural forest and the adjustments nature makes to natural disturbance. For this research the central question was how the forest functions with all its parts. The mechanistic model focused on a managed forest and the products that could be enhanced by controlling certain parts. For this research the central question was how to produce the most timber most efficiently for society. These two approaches lived amicably side by side for decades, but by the 1940s Aldo Leopold, a former Forest Service ranger who later gave eloquent voice to the science of ecology, separated these different approaches as Group A and Group B.

Leopold wrote, "In my own field, forestry, Group A is quite content to grow trees like cabbages, with cellulose as the basic forest commodity. It feels no inhibition against violence; its ideology is agronomic. Group B, on the other hand, sees forestry as fundamentally different from agronomy because it employs natural species, and manages a natural environment rather than creating an artificial one. Group B prefers natural reproduction on principle. It worries on biotic as well as economic grounds about the loss of species . . . It worries about a whole series of secondary forest functions . . . To my mind, Group B feels the stirrings of an ecological conscience."[11]

The stark contrast Leopold drew was remarkably hostile toward forest scientists whose intent was to grow trees as a renewable crop of timber. His description suggested that they were violent and lacked a conscience. Certainly, most people could not be pigeonholed into one group or the other. But differences between ecosystem-based and production-based research would continue to diverge over the next decades, eventually pulling the two types of research apart in a polarizing struggle over science, politics, and ideology in the Douglas fir region.

Leopold's stirring of an ecological conscience grew into a land ethic that inspired later generations of scientists and citizens, but in the 1930s most forest scientists, including T. T. Munger and Leo Isaac, adhered to a utilitarian land ethic focused on sustainable use of the forest. They believed themselves to be ethical people whose research could improve the forest and nature for the public good. They sought knowledge—even ecological knowledge—to help them design a more productive forest. "Without a knowledge of ecology the forester can not hope to purposefully alter the composition of his forest," Munger wrote in 1928. The forest scientist needed to know how nature worked before he could pursue the "wonderful opportunities to improve on Nature, impudent as that may sound."[12]

That impudence might strike a modern ear as arrogant and cavalier, but the idea that nature could be improved was, in the 1930s, not without its own sense of morality. Nature was seen as inefficient and messy, especially in its old forests. E. T. Allen's red fir study and Munger's growth studies had shown that Douglas fir's most productive growth slows after age 80. The New England preacher's son valued thrift and economy. He believed that science could make forests thrifty, economical, and more productive. If he gave slight pause to the thought that improvement could seem impudent, he seemed confident that the improvements he sought could be accomplished with time and experimentation.

Munger's confidence that science could improve nature reflected the spirit of the times. Many of President Roosevelt's New Deal projects, including huge dams on the Columbia River, were rearranging nature on a massive scale. The federal government had never been so involved in so many aspects of American life, putting people to work building, planting, and reorganizing the landscape for greater productivity.[13] Productivity was a good thing, a way out of economic depression, and whatever science could do to improve productivity was believed to be a good thing too.

By the 1930s and 1940s many of the earliest experiments at Wind River were beginning to produce results. In his long-term heredity study, for example, Munger was able to document that high-altitude stock did indeed grow much better than other stock at high altitudes, as did low-altitude stock at low altitudes.[14] And remeasurements by junior forester William Morris confirmed Julius Hofmann's findings that the size and form of seed trees left on cutover lands had no bearing on the "goodness" of the seed. This confirmation seemed to allay concerns over leaving poor, fungal-infected trees as a source of regenerating seed.[15]

Climate data that had been collected at Wind River in handwritten daily measures since 1911 provided a context for understanding long-term studies.[16] (It would be years before computers automated this time-consuming chore.) Time and climate continued to affect the arboretum. Native conifers outgrew the introduced species, even though some larches, giant sequoia, and Port Orford cedar grew well enough to encourage further testing. Eastern hardwoods, which had begun with the most rapid growth at the arboretum, grew so poorly after a few years that no more were added after 1928. However, a few survivors deserve recognition. Several American chestnut trees, once common in the eastern United States and obliterated by an introduced pathogen early in the 20th century, still grow and produce chestnuts in the Wind River Arboretum today.[17]

Controlling fire and revegetating burned land remained utmost concerns to Wind River researchers. One way to keep the forest from burning was to eliminate as much flammable material as possible. Researchers began to fight fire with fire, refining methods of slash burning that would reduce fuels in the woods without burning up the soil necessary to grow the next crop. Using plots that Richard McArdle had established at Wind River in 1926, Isaac measured the amount of flammable material that was left behind after a timber harvest and found that following a typical clearcut, one acre was left with up to 140 cords of wood greater than 3 inches in diameter, up to 200 cords of smaller material, and up to 4 cords of rotten wood.[18] Slash fires following clearcuts consumed mostly smaller material. But what did these slash fires do to the forest soil?

In the first comprehensive study of soils in the Douglas fir region, Isaac and forestry technician Howard Hopkins examined this question. They examined the effects of slash burning on soil following the harvest of a mature Douglas fir stand at Wind River and found that the tried-and-true practice left some sites in poor shape to grow a second crop of trees. Slash burning could destroy the organic matter, often left an ash layer impermeable to moisture, depleted soil nitrogen, broke down the structure of soil, made it vulnerable to leaching, and left the exposed surface dark and vulnerable to solar heating.[19]

Isaac was troubled by the lack of reforestation over 11 million acres of cut or burned forestland in the Douglas fir region. He estimated that at least 50 percent of this area had failed to sufficiently restock and 20 percent had failed to reproduce at all.[20] In response to this and to his findings from the soil study, Isaac compared seedling establishment on burned and unburned portions of several sites within the Wind River valley. He found that slash burning exposed the mineral soil necessary for seedling establishment, but it increased seedling loss due to heat and drought. Following slash burns, the soil had more available nutrients but less organic matter and less capacity to hold moisture, and the sun could make the blackened soil so hot that it cooked new seedlings.[21]

Isaac's advice was to burn lightly where slash was less abundant and not to burn on south-facing slopes, where leftover debris could provide shade. He suggested leaving blocks of uncut timber at intervals that could provide a seed source and aid in fire protection. He further suggested that in areas where stand and site condition permitted, clearcutting should be replaced with shelterwood or selection cutting to reduce seedling mortality.[22] This recommendation would contradict his opposition to proposals about the practice of selection cutting in the Douglas fir region that would surface in a few years.

Speaking to the Ecological Society of America, Isaac acknowledged the ecological role of fire in establishing Douglas fir, confirming what

Allen had observed in 1903 in his report "Red Fir in the Northwest." Historically, fire had created openings where natural stands of single-age, single-species, fast-growing, second-growth Douglas fir had established. Yet settlement and logging had opened far more land than natural seeding could revegetate, making room for weeds. Wildfires that followed slash burns had devastated forest reproduction, killed seed trees, and further receded the timbered edge. Instead of a natural plant succession that would lead to desirable stands of Douglas fir, Isaac warned that much of the cutover forest land in the region was moving toward a permanent plant cover of weeds and brush. Sounding an early alarm against the spread of weedy exotic species such as Scotch broom and Himalayan blackberry into cutover forestlands, Isaac warned that "colonies of exotic species are upsetting natural succession in some localities because they are unpalatable and more vigorous than native species."[23] It would be another 50 years before land managers would take his warning of ecological imbalance seriously.

Munger echoed Isaac's concern about the extent of unreforested land. He reviewed studies at Wind River and elsewhere, tracing the results on forestland that had received different treatments of slash disposal and burning. He concluded that repeated fire was the culprit, carried by two features that Munger considered villainous: slash and snags. Munger outlined advice for forest landowners on how to burn slash, when to burn, and where and where not to burn. Free from risk, these harvested lands could be safely reforested, or, Munger noted, they could be kept open for incidental commercial uses such as berry picking, cascara bark collecting, or beekeeping.[24]

Despite the ecological warnings about slash burning, the practice continued to be a necessary part of what was thought to be a well-managed forest, if for no reason other than to control wildfire. A six-year study by Morris at Wind River, H. J. Andrews, and Cascade Head experimental forests concluded that "although burning slash on most patch cuttings similar to those studies does not decisively reduce or increase numbers of postlogging seedlings, it does facilitate fire control."[25]

Although slash burning may have hindered natural seeding, it certainly made it easier to plant with seedlings. Increasingly, reforestation relied less on natural seeding and more on planting, so slash burning continued as a forest practice through most of the 20th century. But there were still problems with the method. Concern over air pollution and the occasional runaway fire would result in increasing management costs and ecological consequences. After his retirement, Isaac described the complex relationship of fire to the development of Douglas fir forests:

> Our beautiful virgin forests of Douglas fir followed fire. Since the advent of cutting and logging, burning has been found to be not always indispensable but often a useful tool in eliminating fire hazard and preparing a site for a new stand. Foresters have learned under what conditions its use may be beneficial, and where its use may be harmful. It has also become evident that there is still much to learn about the biological and physical effects of burning in this forest type. Therefore, ecologists and foresters must hesitate a long while before they propose burning as a blanket rule or before they try to eliminate burning as a prime factor in the ecology of this type.[26]

More than any other researcher in his time, Isaac understood the ecology and management of Douglas fir. Quite literally, he wrote the book. Taking off from Hofmann's initial seed studies, Isaac explored Douglas fir silviculture from every angle and with a rigor previously unseen in forest experiments. Only after careful research design and years of study was Isaac willing to publish his findings. That is why there is a lag of nearly a decade between the time Isaac came to Wind River and the time his research findings began to see their way into print. After almost 20 years of research, in 1943 Isaac published what is still considered the definitive work on Douglas fir. In *The Reproductive Habits of Douglas Fir* Isaac synthesized results of past research to provide foresters with practical information "for continuous production" of Douglas fir. W. B. Greeley, who had been chief of the Forest Service from 1920 to 1928 before joining the timber industry, wrote in the publication's foreword, "As to this book on growing Douglas fir, it should be in the hands of every forest owner in the Pacific Northwest."[27]

In his report, Isaac detailed the complexity of Douglas fir—the variability of seed crops, soils and climate, problems with rodents, and vegetative competition. He recognized the severe ecological consequences of unregulated clearcutting, including increased erosion and vast acreages of unregenerated forestland. He urged consideration of local environmental factors before applying management prescriptions and warned against simple rules applied too broadly. "Conditions within the Douglas fir type are so varied that simple, specific rules cannot be set forth for securing regeneration over large areas," he concluded.[28]

Yet the simple, specific rule of clearcutting followed by slash burning and replanting had become well-established by the 1930s. Each part of the three-step prescription was rooted in accepted scientific understanding of how Douglas fir naturally developed. Clearcutting was considered to be a mechanized substitute for the natural role of forest fire, clearing land for a flush of new growth. Burning slash eliminated competition from undesirable hemlock, reduced fire hazards, and exposed the mineral soil that Douglas fir seeds seemed to naturally favor. Planting ensured reforestation with the most desirable species. It was a prescription meant to increase the forest's natural potential for producing timber. It was simple, science-based, and widely applied.[29]

In addition to this scientific rationale, Forest Service economist and forest supervisor Burt Kirkland outlined an economic rationale for clearcutting in 1911. Kirkland's argument centered on the concept of sustained yield, an idea adopted from European forestry, where timber harvest equaled growth. Timber was money in the bank, but old timber gained no interest. In the national forests of western Washington and Oregon, much of the timber was old. According to Kirkland, decadent trees grew only more decadent, and their growth was more than offset by decay. These stands offered no benefit for the future, and he asserted that saving them would be uneconomical.[30]

Selective cutting was practiced in Europe and in eastern American forests to cull decadent trees. Kirkland advocated clearcutting in the Douglas fir region. Old Douglas fir trees were huge and heavy. Kirkland maintained that logging old growth in the steep terrain of the Pacific Northwest meant that younger trees "must inevitably be crushed by felling the old timber and by dragging out the heavy logs." And because younger trees that had grown in the shade of older Douglas firs "consist almost entirely of inferior species," the forest that would remain after selective logging would be "only a partial stocking of very inferior trees."[31]

With scientific, economic, and technological arguments to support it, clearcutting seemed like a rational management practice for the Douglas fir region. It also created new problems in the form of vast areas of cleared land that needed to be replanted. Although forest policy up to this point had not required reforestation following clearcutting on private forest land, the collapse of the world economy in the 1930s created a more convincing context for such regulation. By that time, research at Wind River was moving away from the natural history study of silvics and toward its application in silviculture. Scientists were less concerned with the forest that existed in nature and more concerned with the forest that could be created or improved.

To "improve" the forest meant to replace its decadent old growth with faster-growing trees that held more promise for profits in the future. In one study in 1936, Isaac established 23 one-acre plots to study stand improvement in parts of the old-growth forest where decay seemed to far outpace growth. He girdled old, defective trees to see if the stand could be "restored" and net growth quickened by younger trees. Although "defective" was not defined, trees that were labeled as such probably included those with conks, rot, or broken tops. Stand improvement studies continued throughout the middle of the 20th century in an ongoing attempt to create "improved" forests with a minimum of decay, pests, diseases, or undesirable species. At that time, biological diversity was something to eliminate, and clearcutting was a good tool to do the job.[32]

Although embraced as a pragmatic approach to improving forests in the Douglas fir region, clearcutting was less popular in other parts of the country. In 1903 controversy erupted at the New York School of Forestry, founded by Bernhard Fernow, because of an uproar over silvicultural experiments that involved clearcutting. Fernow recalled that although the school was thriving, with enrollments that surpassed any French or German forest school at the time, funding was suddenly pulled by the state legislature in 1903 because of clearcutting in the demonstration forest, and the college was forced to close.[33]

Undeniably, clearcutting was a hot-button issue. When President Roosevelt toured the Pacific Northwest in the 1930s, he and his entourage were reported to have objected to the unsightly view of clearcuts.[34] Although scientists could provide a pragmatic explanation for the view, they could not excuse it. But the economic engine and demand for lumber that had encouraged ever larger and unsightly clearcuts had suddenly ground to a halt by the time President Roosevelt visited the region. The timber industry was failing. According to Richard Rajala, it was during this time that industry officials began to consider an alternative to clearcutting in the Pacific Northwest. Instead of glutting an already depressed market with more unwanted lumber from clearcutting, the industry turned to logging only the best trees and milling only the best parts of the best tree.[35]

This was not a return to horse logging and bull teams, but rather the application of a newly developed technology: the 80-horsepower diesel tractor. According to PNW forest economist and engineer Axel Brandstrom, the tractor could negotiate rough terrain and harvest trees without leveling the landscape as the old high-lead system had done as it dragged heavy logs across the ground to a central point. Tractors made it possible to remove individual or small groups of trees while leaving surrounding trees untouched to continue to grow for future harvest. Such new, motorized, mobile logging equipment was adaptable to regional conditions, and Brandstrom argued that it could put an end to large-scale clearcutting.[36]

Obviously impressed by the argument, Kirkland collaborated with Brandstrom in 1931 to find ways to make the long-depressed lumber industry more profitable.[37] Twenty years earlier, Kirkland had supported clearcutting on both economic and technological grounds. Now he argued that rebuilding a forest from scratch following clearcutting involved investment and accumulated costs that were compounded over a long time and fraught with uncertainties. According to Kirkland, selective timber management would provide an immediate return as well as a continuous future return from timber of all sizes, from fence posts to logs measuring up to 40 inches or more. Clearcutting, in contrast, provided no possibility for income for a generation or more.[38]

Kirkland and Brandstrom set out to change forest management in the Douglas fir region with a design that they argued mimicked nature's pattern of growth, maturity, and decline. By 1936 they had completed a study that recommended harvesting small groups of trees in 2- to 10-acre patches, just big enough to expose mineral soil to sunlight and small enough to allow the forest edge to supply seed for natural regeneration. Unlike clearcutting, this system would provide an abundant seed source, an unaltered forest climate, and a reduced threat of fire. The silvicultural emphasis would switch from regeneration of single-age stands, which had been unsuccessful in many places throughout the region, "to encourage where feasible the perpetuation of the mixed forest as better fitted to meet the industrial requirements of the region than a pure Douglas fir forest. The mixed forest is also universally recognized as the safest from insects and disease."[39] Kirkland and Brandstrom called their design "selective timber management," and it was meant to provide a continuous supply of timber from a forest of mixed ages and species, with concern for forest aesthetics and sustainable timber supply.

They would meet great resistance from the forest researchers at Wind River. Part of the problem stemmed from a confusion between selective timber management and selective logging, or high-grading. A definition from Raphael Zon's *1927 Yearbook of Agriculture* defined selective logging as removing the largest and most valuable trees and leaving the smaller and less valuable trees. This sort of high-grading reflected Kirkland's original arguments in favor of clearcutting. But the selective timber management that Kirkland and Brandstrom proposed was nothing like the selective logging that Zon had described nor the high-grading that Kirkland had warned against. Their selective timber management considered the hierarchy of trees in a stand and stands in a landscape, "and at no time are operations based on uncertain predictions of distant future rates of growth." They advocated a system of continuing experimentation that would be flexible, in order to provide "aesthetic, protective, and other functions of the forest."[40] Their concepts of adaptive management, forest aesthetics, and silvicultural systems other than clearcutting would not be fully tested in the Douglas fir region for another half century.

In 1936 the idea to selectively manage Douglas fir forests found a sharply divided audience. Although the proposal had strong support from both the Washington and Region Six offices of the Forest

Service, it drew fire from Munger and Isaac, those who arguably knew the most about growing Douglas fir.[41] However, before Kirkland and Brandstrom published their report, both Munger and Isaac had supported selective cutting. In 1933 Munger wrote, "In the opinion of the writer, the guiding principle in the silviculture of over-mature stands of the Pacific slope should be selective in cutting . . . The applicability of partial cutting to each type and set of conditions should be thoroughly tested."[42] In 1930 Isaac argued that selective or shelterwood cutting could replace clearcutting in some cases, greatly reducing seedling losses from heat and drought and providing a most abundant seed supply, assuring a more rapid rate of restocking following logging.[43] Despite Isaac's warnings against blanket prescriptions and Munger's call for thorough testing of partial cutting, the two scientists seemed to be unequivocally against Kirkland and Brandstrom's idea of selective timber management.

In retrospect, the arguments that Munger and Isaac posed against selective timber management seem narrowly focused, almost narrow-minded. Although the report made a clear distinction between the proposed system of selective timber management and potentially destructive types of high-grading, both Munger and Isaac argued against it as if it were high-grading.

"The tendency was to high-grade, take the high value trees and leave the lower grade ones or the lower grade species," Munger recalled years later, "take the Douglas fir and leave the hemlock, which was a high-grading practice; that's what this selective cutting was inclined to degenerate into. It had the immediate economic advantage of harvesting much of the value but leaving a forest in a rather deplorable condition, if practiced in the extreme." Recalling a German play on words, Munger admonished that a "plenterwald" (selection forest) must not become a "plunderwald" (plunder forest).[44]

Isaac also focused on high-grading rather than on the proposal at hand. "So after a couple of cuts the firs would be gone and then you had nothing but these secondary trees coming along . . . It's like using mares for milk stock on a dairy farm. It's just not the right animal."[45]

Another problem may have been that Kirkland and Brandstrom's report was created beyond the control of Munger's influence. Kirkland and Brandstrom had been assigned to Munger's experiment station by

Ray Marsh, assistant chief of the Forest Service in the Washington office. "Ray Marsh hired every crackpot that showed up on the horizon in the Forest Service," Isaac recalled. "Here in our midst were two fellows with crackpot ideas. [Marsh] got a hold of them and hired them in spite of opposition from the rest of the Service. Then C. J. Buck, the regional forester, got on the bandwagon mostly to spite Munger, we thought . . . Munger opposed their coming, but they were assigned anyway and given more or less a free hand."[46]

Kirkland and Brandstrom worked at the station for five years, developing their proposal for selective timber management. Munger and others reviewed the final manuscript prior to publication and sent a lengthy response to the Washington office listing all their objections. With their objections overturned, Munger traveled to Washington, D.C., to personally deliver his case against the report. After extensive negotiations, Kirkland and Brandstrom's report was published not as a station report but as a special publication of the Charles Lathrop Pack Forestry Foundation, a nonprofit organization independent of the Forest Service.[47]

Debate continued after publication. Some saw the report as the ultimate management for the Douglas fir forest and others saw it as the forest's ultimate demise. "But there was bad blood between Munger and C. J. Buck because of their differences," Isaac noted years later. "And C. J. Buck used this opportunity to thwart Munger by putting selective logging into effect on the national forests."[48]

Presumably to test the concept of selective timber management, Isaac examined stands at Wind River and in the national forests where selective cutting of Douglas fir had occurred in the past. Again, what he tested was high-grading, not the strategies outlined by Kirkland and Brandstrom. He examined only stands where selective cutting of individual trees was done and where the trees removed were all of high quality. He sampled across the region with no attempt to randomly select sites or establish control sites. Twenty years later, in 1956, he published his findings debunking selective cutting in the Douglas fir region. But the report did not reflect the same rigor Isaac had demonstrated in other studies.

Isaac reported of his 10-year study, "No effort was made to get a true random sample of entire cutting areas . . . check areas could not be set up for statistical comparison of growth and mortality in uncut and partially cut areas . . . Instead, the areas were treated as case studies, the results averaged, and general conclusions drawn from these averages."[49]

The new system that Kirkland and Brandstrom had proposed was never truly tested. For the next 40 years, forest research looked almost exclusively at clearcutting and not at other systems. Knowledge that might have been gained from thoroughly testing Kirkland and Brandstrom's proposals, doing the adaptive management that they advocated, and adapting findings to specific sites and needs would have been most helpful at the end of the 20th century, when forest managers would be required to manage forests for aesthetics, recreation, and biological and structural diversity as well as for timber. According to Robert Curtis, PNW research mensurationist, "An unfortunate result of this episode was the abandonment of efforts to develop alternative silvicultural systems," efforts that would not be revived for another 50 years.[50]

The controversy surrounding the selective timber management proposal of 1936 underscored the sometimes uncomfortable fit between policy, management, and science, and the distrust that underlies assumptions made in the course of research. Munger and Isaac were so distrustful of how selective cutting could be misapplied that they never fairly tested Kirkland and Brandstrom's proposal. Their distrust led to a polarized debate that did not allow for a full range of options for managing forests, and limited the tools and knowledge available to forest managers. Similar discomfort and distrust would surface again 50 years later, when scientists would clash in the debates over old-growth forests in the Douglas fir region.

When Munger began his career in the 1908, forest scientists worked with considerable freedom to devote almost exclusive attention to their research. As research programs grew, so did the administrative burden. By the mid-1930s, meetings, conferences, and a steady stream of professional visitors from abroad occupied much of Munger's time, and the Washington office preempted much of Munger's decision-making authority. In 1938, after years of escalating responsibility, Munger took ill. For four months, he was on sick leave. Shortly after returning to work, Munger relinquished his position as director of the PNW Research Station to resume full-time research.

As it turned out, the policy of selective timber management was short-lived. As bombs fell in the Pacific, bringing the United States into World War II, forest research shifted toward yet another war effort, but that war effort found much of the private timberland already cut. Between 1939 and 1942, as sawmills turned to the federal forests to provide more raw materials, harvests in the national forests increased sevenfold, with most of the cutting in the Pacific Northwest.[51] Since its earliest days, the timber industry in the Pacific Northwest had suffered cycles of boom and bust. And for almost as long, there had been warnings coming from the industry and the agency about the need to manage the nation's timber on a sustainable basis and stabilize forest-dependent communities. The idea of sustained-yield timber production was put forward as early as the 1920s as a way to limit the volume of timber on a glutted market through cooperatively managed units of private and public timberland. But with the nation at war again, timber production geared up and did not slow down for 40 years. Finding ways to maximize that productivity would dominate forest research in the Douglas fir region for most of those years.

Chapter Six

Research Expands Following the Second World War

✣

Large, continuous bodies of heavy timber are virtually biological deserts.

—Leo Isaac[1]

World War II changed everything. The nation's war effort brought an influx of people and industry to the Pacific Northwest and took out record-breaking volumes of timber. War-related labor shortages encouraged the timber industry to adopt more mechanized production processes, including the use of gas-powered tractors and chain saws. The victory that followed the atomic bombings of Japan had proven the power of science and technology. Now in peacetime, the nation's science leaders saw the promise of full employment, health, and prosperity "if we make use of our scientific resources."[2]

The reordering of nature that had begun before the war continued after the war in an exuberance of technological innovation that promised a better life and a more prosperous future for every American. In 1945 President Truman called for the construction of up to 1.5 million new homes to be built each year for the next 10 years.[3] As the nation moved forward toward a new, expansive image, optimism and innovation defined the timber industry of the Pacific Northwest. In the 1950s, timber harvest from the national forests increased from 3.5 billion to 9.3 billion board feet, with a third of that coming from the Pacific Northwest.[4]

Up until World War II, much of the forest research at Wind River focused on controlling fire and understanding natural regeneration. Following the war, research turned increasingly to ways to maximize the production of timber in managed plantations. Timber harvests increased each year following the end of the war; then in 1950 the Korean War bumped demand even further. By the early 1950s, according to historian David Clary, any ambitious young forester could see that in the Forest Service the timber program was where careers were to be made.[5] There seemed to be no limits on the expanding market for forest products at home and abroad. Richard McArdle, who served as Forest Service chief from 1952 to 1962, estimated in 1958 that 24 billion board feet would be required annually from the national forests by the year 2000.[6]

The nation's new hunger for building material silenced any remaining debate over selective timber management. Clearcutting and planting fast-growing Douglas fir became the prevailing silvicultural prescription for most of the forestland in western Washington and Oregon. Long-standing observations that Douglas fir seed would germinate most successfully on cleared land had led to the practice of clearcutting, which continued even after natural reproduction from seed had been replaced by planting with selected seedlings. Clearcutting was a cheaper and more efficient way to manage the forest for timber and to replace slow-growing old-growth forests with fast-growing managed plantations. Increased cutting on the national forests cleared the way—and the forest—for extensive new timber plantations and helped feed the nation's hunger for building material.

Lines began to blur between experimental research and production of timber as pressure grew for Wind River to contribute to escalating national forest harvests.[7] "Speeding up the transition of wild forests to thrifty managed stands" was the overall objective for Wind River and other experimental forests, as outlined by PNW forester Phil Briegleb, who would be the PNW Research Station director between 1963 and 1971. To that end, Briegleb recommended a "major harvest on the main division of the Wind River Experimental Forest, with very careful follow-up regeneration . . . with genetically improved stock."[8]

While Briegleb was drafting his research plan for the experimental forests in the winter of 1950, snow began to fall at Wind River. In an average year, almost seven feet of snow falls at the Wind River Nursery and average winter temperatures hover around freezing. But the snow of 1950 was memorable, breaking all records that had been kept at the station since 1911. By the end of January, 13 feet of snow had fallen and the average temperature had not risen above 19°F. With roads blocked, Wind River residents with cabin fever used a bulldozer to plow an 11-mile path just to get to the town of Carson (fig. 6.1).[9]

The heavy, wet snow challenged trees as well as people. In the arboretum, larches and pines that had shown promising growth snapped from the weight of the snow. Trees in the heredity study were damaged by top freeze. Nature continued to challenge Wind River the following spring and summer, when a severe drought hit the region. Autumn followed with strong winds that toppled trees throughout the western Cascades. Trees that had been damaged by ice, drought, and winds were host to severe outbreaks of bark beetles in the following few years. Another autumn cold spell hit a few years later when fall temperatures plummeted suddenly to below freezing and remained there for six days. The sudden cold affected the arboretum's collection of trees in different ways. Most native conifers were spared damage from the freeze because, accustomed to summer drought, they had hardened off weeks earlier. But the nonnative pines—Coulter, Apache, knobcone, and shortleaf—were still growing vigorously through the

Fig. 6.1. The Regional Training Center buried in snow during the winter of 1950. (Courtesy of USDA Forest Service.)

mild fall and were killed outright by the sudden freeze. Robert Steele, a forester at Wind River during this time, observed that "introduced species may thrive for many years, but severe frost damage or complete loss is an ever-present peril."[10]

It was not only the weather that introduced new challenges to researchers at Wind River, but the focus of research itself was changing as well. Forest research was expanding beyond Wind River to the H. J. Andrews Experimental Forest and several new experimental forests on private industrial timberland where cooperative research could focus on managing young plantations. Wind River was no longer central to research in the Douglas fir region, as Leo Isaac and T. T. Munger began wrapping up their long and productive careers.

Before he retired in 1948, Munger remeasured the permanent plots he had been studying since before World War I. He concluded that yield would be much greater if young trees were thinned prior to harvest, rather than risking the loss of so many trees to crowding, wind, and insects.[11] Munger's study would be revisited in the coming decades as scientists tested methods to increase timber production. Munger also took a final look at ponderosa pine, the tree that had first drawn him to the West in 1908. Very early in his research, he had recognized great variety in the form and growth of ponderosa pine throughout its range, and in 1926 he had sown seeds from 10 races throughout the West at established plantations at Wind River and other sites, both arid and rainy. In 1947 he reviewed the results after 21 years of growth and found that the well-watered sites with long growing seasons generally had better results than drier plantations. Yet it seemed to Munger that other factors might be affecting the pines' growth and mortality.[12]

At midcentury another Wind River pioneer was ending a distinguished career. In the years just after World War II, Isaac, ever the careful researcher, felt confident enough to publish a retrospective of all that was known about Douglas fir seed. Isaac took a six-month leave from his regular duties to compile his report *Better Douglas Fir Forests from Better Seeds*.[13] Due in large measure to Isaac's detailed work, Douglas fir had become a staple of reforestation in temperate zones around the world. In this summary report, Isaac correlated seed sources and climate zones to predict the success of particular races of trees in regions around the world. Drawing on evidence from the Wind River heredity study and studies at the Wind River Arboretum, he described two steps toward a better, more profitable crop of Douglas fir trees: the use of local seed and the gradual improvement of the species "through selection, hybridization, change in number of chromosomes, or use of growth hormones."[14]

Isaac's report would stand the test of time. More than 50 years after its publication, it is still a reliable source of information about the silvics of Douglas fir. Isaac's vision of the gradual improvement of the species would not be realized for many decades. Although much of Isaac's career had been spent understanding the complexity of natural regeneration, by his retirement in 1956, he acknowledged the increasing emphasis to improve upon nature. "The let-nature-take-its-course attitude has been all too prevalent in the provisions for Douglas fir natural regeneration," Isaac said. "In actual practice, we have just learned to manage the tree in its natural wild state, and have not even learned to select the best species or strain for propagating forests tomorrow." In the future, he said, "the forest that is produced [will be] from the stock the [nurseryman] raises."[15]

Perhaps the most lasting legacy of both Munger and Isaac was the next generation of forest scientists they inspired. Following World War II, with the new GI bill, forestry schools were sending out brand-new forest scientists in a wave of professionalism similar to that of the first graduates from the earliest forestry schools at the beginning of the 20th century. Of the new generation of forest researchers, forest geneticist Roy Silen worked most closely with the older scientists, particularly with Isaac. Silen was hired as Isaac's assistant in 1946, and he remembered the older scientist as a careful observer and researcher. "Everything in his book *Reproductive Habits of Douglas Fir* was essentially correct, even as far as we could tell years later. So, this was an unusual experience to be the assistant to the person who knew more about Douglas fir silviculture than anybody."[16]

Silen took on the stewardship of the permanent sample plots, remeasuring the arboretum, the heredity study plots, and the Wind River transect. He continued monitoring the Munger's ponderosa study after Munger's retirement.[17] Not long before his own death in 2004, Silen revisited the Wind River ponderosa pine study and the others in Arizona, Idaho, and New Zealand. He concluded that, as had been found with Douglas fir, after 50 years native ponderosa pine outgrew all other racial groups in each area, not so much in height growth as in mortality rates. The value of these old and continuing studies was underscored in Silen's final report. "Time has no substitute . . . Only long-term studies can fully evaluate outcomes of cumulative stresses of such long duration, as trees, like people, also age, weaken and fail," he wrote.[18]

Silen forged the strongest link connecting the long-term studies begun by an older generation and the new research agenda that emerged following the war. His dedication kept many of the long-term plots measured and recorded throughout a time when such work would be seen as less than vital to forest science. And he would in turn inspire yet another group of scientists coming to the forest in the 1960s and 1970s.

Steele and silviculturist William Stein also came to Wind River after the war and established permanent sample plots in the Wind River Research Natural Area and expanded the spacing study Isaac had established in 1925. They examined the incremental effects of pruning and thinning done earlier by the Civilian Conservation Corps (CCC) and monitored the weather station. But not all the work of these young scientists was scientific. Stein recalled, "At noontime, we would sit on a bridge over the Wind River . . . and dropped grasshoppers and watched the trout. It was great fun."[19] The young scientists served as ambassadors to visiting foresters who came to Wind River from around the world to learn techniques for growing Douglas fir. Stein and Silen tromped visitors through the arboretum and up to the spacing study, where they showcased experiments that had helped launch Douglas fir as the tree of choice for reforesting much of the world's temperate zone. There was much to show for 40 years of research, and new questions were emerging.

One new area of emphasis was forest soil. In 1946 soil scientist Robert Tarrant transferred from the Soil Conservation Service to begin soil studies in the Douglas fir region and to head the administration of the Wind River Experimental Forest. Although he later became the director of the PNW Research Station and his fieldwork stretched across the region, Tarrant held a special affinity for Wind River. "It was the site of so much of the first work, the original studies that defined the region, a place of important discovery," he recalled years later.[20]

Tarrant made one of those important discoveries near Planting Creek in 1948. While making a soil reconnaissance on a hilly slope on a hot day "through a nondescript, yellow-foliaged Douglas fir plantation," he unexpectedly came upon a shady forest of red alder mixed with Douglas fir. "Perhaps because the day was hot and the mixed forest was unusually cool, I stayed there longer than I might have otherwise," Tarrant remembered.[21] He spent several hours wandering through the alder and Douglas fir, where it was several degrees cooler than the surrounding plantation. He noticed that the foliage of the Douglas fir planted among the alder was exceptionally dark green, the needles were more abundant and larger, and the surface of the forest floor was thick with leafy duff.

Back at the office, Tarrant asked Isaac what he knew about the mixed forest he had stumbled across. Isaac recalled a story from 1929, when he was working with a tree-planting crew following a fire in that part of the forest. During a lunch break, one of the tree planters said that he would have bet five dollars the fire would have stopped right there on the ridge if there had been a stand of alder there. Isaac had been intrigued; he arranged to have some red alder seedlings grown in the Wind River Nursery. Serendipity intervened as Isaac's alder seed was slow in arriving, which gave the Douglas fir seedlings a head start on the faster growing hardwoods. Eventually the alder seedlings were planted and seemingly forgotten by all but Isaac. "Obviously, few people knew about the plantation, and fewer yet had any idea of the lessons it held for forest managers," Tarrant said.[22]

But the lessons did not escape Tarrant. Subsequent studies of the alder strip showed that Douglas fir within the mixed stand were larger than those in the pure stand, with more than twice the total volume of wood and more nitrogen in the soil than on the pure Douglas fir plantation.[23] Tarrant's initial observation led to more detailed study of the benefit of red alder, which up until then had been considered a brushy nuisance. Tarrant's studies confirmed that alder increased nitrogen and organic matter in the soil, benefiting the subsequent establishment and growth of Douglas fir.[24]

Many forest soils within the Douglas fir region suffer from inadequate amounts of nitrogen, slowing potential tree growth. Interest in nitrogen-fixing alder soon led to interest in chemical fertilizers. The leap from natural to chemical fertilizers reflected an assumption that characterized the years following World War II: technology could improve whatever nature provided. Another young silviculturist, Donald Reukema, examined the effect of ammonium nitrate on a low-quality site on the same plantation that held the alder strip. In just four years, Reukema noted that height and diameter growth of the fertilized Douglas fir trees had surpassed that of the unfertilized trees.[25] Soil scientist Richard Miller joined Reukema in a 1970 study that added nitrogen fertilizer to individual, dominant Douglas fir trees on a thinned stand growing on another poor site in the experimental forest. The 20-year-old trees increased average diameter and height growth by 85 percent over the unfertilized trees.[26] Although these findings encouraged the use of chemical fertilizer on low-quality sites and recently thinned stands, results from similar experiments on higher-quality sites in the region were less dramatic. Tarrant recognized that chemicals would continue to have a role in forestry but called for "the ecological knowledge necessary to manage and maintain a healthy biosphere with minimum use of chemical tools."[27]

A subtle shift in research emphasis was occurring at Wind River by the mid-20th century. The physics and behavior of fire, which had been the focus of much research by Julius Hofmann and others at Wind River, were being studied by researchers stationed in the more combustible forests of Idaho and Montana. The natural patterns of growth and reproduction of Douglas fir, a topic that Munger and Isaac had spent their careers researching, was largely replaced by new research focused on maximizing growth and controlling whatever interfered with optimum production. There was a new assumption underlying forest research at that time, an assumption that nature could be controlled and thereby improved. The work of the first generation of scientists at Wind River had demonstrated that Douglas fir could be grown as a crop. Now it was the next generation that would find ways to improve the crop and protect it from threats of disease, pests, fire, and competition from less valuable trees and marauding wildlife. This tangle of forest species and natural disturbances that characterized the wild forest was, at the time, considered messy at best and threatening at worst.

Mice, for example, seemed to be taking more than their share of tree seed, now that seeds were being broadcast from airplanes onto ever-larger expanses of clearcuts. Wildlife biologist Donald Spencer arrived at Wind River to field-test a special poison developed to kill rodents. In a study dubbed "Operation Peromyscus," Spencer measured mortality and reinvasion rates of mice and how well the poisoned seed weathered the Wind River climate. But the seed poisoned other animals, including birds, and Spencer conceded that mice had been consuming an abundance of insects, which were even more destructive to trees than mice. Despite these findings, the study forged ahead with poisoning hundreds of acres of cutover land.[28]

In the early 1950s ice, wind, and drought damaged thousands of acres of forest, and infestations of bark beetles threatened the loss of more than 2 billion board feet of timber throughout western Washington and Oregon. The immediate question for public and private foresters was how to salvage all this timber before it rotted. By now some larger timber companies were establishing research programs of their own. Collaborating with scientists in Weyerhaeuser's new research department, Wind River researchers found that rot advanced most rapidly during the warm season. They found that sapwood deteriorated quickly and rotting trees splintered when harvested. The higher mortality increased the risk of fire. Based on these findings, the scientists concluded that to maximize production and minimize fire danger, damaged trees should be salvaged promptly.[29] Another scourge that threatened Douglas fir was root diseases. Thomas Childs, a forest pathologist, found root rot in a young stand at Martha Creek, and established plots there and in the Panther Creek Division, where he studied the disease progression for more than 25 years. He found that new infections could result from contact with the root systems of infected trees, and warned that if control measures were not taken after the old stand was harvested, infection could spread to future stands.[30]

As many of these new studies demonstrated, timber production was the overriding measure of a forest's productivity after World War II. The agrarian model of forestry that encouraged growing timber as a sustainable crop was becoming more technological. Good stewardship in the forest, like good husbandry on the farm, was being driven by more sophisticated techniques that would dramatically increase productivity. The Douglas fir region was being converted from forests of many different species and ages to plantations of single-species, single-aged monocultures of Douglas fir. The goal of forest science was to make these plantations produce timber efficiently, with the greatest possible growth in the shortest possible time.

As the nation became obsessed with its technological future, intensive forest management promised that by applying technological know-how, trees of the future would grow faster than trees of the past. And faster growing trees meant larger returns on the investment and more homes for a fast-growing nation.[31] Airplanes were used for aerial forest surveys, locating concentrations of forest pests or disease, and delivering chemical treatments to infected areas. New chemical herbicides, insecticides, and rodent killers promised more broad-spectrum control with their ability to kill many kinds of organisms that threatened timber production. Certainly there had been technological approaches to most human industry throughout time, from stone tools to satellites. But during the postwar boom, technology itself became a focus of research, and growing the best trees became a technological challenge.

For the most part, these changes were seen as positive steps toward the goal of efficiency and productivity, overcoming the uncertainties and wastefulness of untamed nature. After the war, the population of the Pacific Northwest ballooned, especially in cities where the Columbia River's new hydroelectric dams powered a burgeoning defense industry. As historian William Robbins pointed out, the basic science that preceded the war and the technological explosion that followed created the conditions for a massive reordering of nature.[32] A belief in American technology built the great dams of the West and would soon launch the Space Age. Technology fueled an American race toward progress and prosperity and seemed to make it possible to improve nature on an epic scale. Timber harvest and production continued to accelerate. Most people believed that this acceleration contributed to a better, more prosperous nation and that the Forest Service's role in that prosperity was to be admired.

Smokey the Bear, the Forest Service's postwar poster image, peered stolidly from the cover of *Newsweek* on June 2, 1952. "No one can deny that the Forest Service is one of Uncle Sam's soundest and most businesslike investments," the magazine article proclaimed. "It is the only major government branch showing a cash profit and a growing inventory. This year, through timber sales, grazing permits, and other fees, the foresters will turn back to the U.S. Treasury a net surplus of $10 million . . . The annual value of free recreation, wildlife management, increase in the value of timber stands, and particularly the pure and abundant water easily tops half a billion dollars." The magazine credited the agency's success to a spirit of responsibility and loyalty among its ranks. "The man who rises through the Forest Service is of a peculiar breed. He is a woodsman, a scientist, an engineer, an economist, an accountant, a public-relations expert, and something of a nomad." After a half century, those in the ranks of the Forest Service still reflected the image of Gifford Pinchot.[33]

The Columbia National Forest in southwest Washington was said to have been Pinchot's favorite of the national forests. Following his death in 1946, the forest was renamed the Gifford Pinchot National Forest in honor of the Forest Service's first chief. A commemorative ceremony in 1949 celebrated the man and the forest he loved, a place of great beauty and high economic value. At the time, managers estimated that the forest could produce a sustained yield of 200 million board feet of timber each year. Ten years later, the annual harvest from the Gifford Pinchot National Forest was almost 262 million board feet, and by 1968 the harvest level was more than 550 million board feet.[34] A board foot is a volume of wood measuring 1 foot by 1 foot by 1 inch, and 550 million board feet is enough wood to build 50,000 two-bedroom houses. The United States continued its building boom through most of the 1960s and 1970s, and the forests of the Douglas fir region in western Washington and Oregon were the prime supplier of lumber.

Chapter Seven

Better Timber Faster

The interpretation of results at 50 years is almost diametrically opposed to interpretation of results at age 20.

—Roy Silen[1]

In 1962, as world powers faced off over missiles in Cuba and John Glenn orbited Earth in a space capsule, Rachel Carson published her book *Silent Spring*. In it she warned of the dangerous and unintended consequences of chemical pesticides on wildlife and humans. Other negative consequences of America's technological prosperity, including a massive oil spill in the Santa Barbara channel and the detection of DDT in human breast milk, began to seep into the public consciousness during this time. Technological control of nature would deliver more unintentional consequences in the years to come.

Despite the public stirrings of an ecological consciousness, a can-do spirit characterized the Forest Service and its research support of intensive timber production. Since Gifford Pinchot's time, the Forest Service had worked to accommodate multiple uses within the nation's forests. The two uses initially outlined in the Organic Act of 1897 were timber production and water supply. The concept expanded with the Multiple Use–Sustained Yield Act of 1960 to include forage, recreation, and habitat for fish and wildlife. Soon wilderness was added to the escalating concept of multiple use. But timber production came first, especially in the Douglas fir region, in the 1960s. With emphasis on intensive plantation forestry, work at Wind River increasingly focused on silvicultural methods to speed the growth of trees from seedling to harvested log.

Many of the studies established in the early years at Wind River would reveal surprising results in the second half of the 20th century. Donald Reukema, who joined the PNW Research Station in 1953, used many of Wind River's old plots to examine ways to coax a higher yield from young stands of Douglas fir. For 30 years, Reukema examined the set of precommercial thinning plots first established by Julius Hofmann in 1919 and expanded by Walter Meyer in 1934. At one point measurements showed that the twice-thinned plots produced greater yields, but later measurements showed little difference between the thinning strategies. Reviewing data that spanned more than 60 years, Reukema eventually concluded that none of the plots had reached the point of maximum annual growth, and that any difference in yield among them appeared to be related to site quality rather than thinning treatment. Reukema's study demonstrated that long-term trends in forest growth would eventually reflect environmental variations in sites and plots, making it difficult to predict the effects of particular treatments.[2]

The thinning study illustrated the trouble with field experiments. Unlike laboratory studies, where something as variable as site quality could be standardized, working in the woods presented scientists with unlimited variations. To reduce the possibility of false conclusions, experiments had to be repeated in every detail at other sites in a process called replication. It was difficult—impossible, really—to find sites with identical soil, climate, and fire history. Nature was not standard. But without replication, it was too easy to misinterpret findings.

Reukema continued to examine ways to increase timber production on tree plantations. He followed up the spacing study that Leo Isaac had established in 1925 to test how planting distances would affect tree growth. Reukema found that after more than 50 years, trees planted at closer spacings (4 and 6 feet apart) had less total wood volume by plot, and that the individual trees were slow to thin themselves, suffered more mortality, and produced less merchantable timber. He found that trees planted at wider spacings (8 and 12 feet apart) continued to grow taller, with larger diameters and a greater volume of merchantable timber. But most surprisingly, Reukema often found greater variations in growth among trees within the same plots than among trees in plots of different spacings.[3] There were places where big trees grew close together and smaller trees grew widely spaced apart. These variations among the trees in Isaac's spacing study intrigued many scientists, including Richard Miller, a PNW Research Station soil scientist. Miller questioned whether such variation could be attributed to soil quality rather than competition for space. Four samples taken in 1961 indicated that the soil was of about the same quality across all the plots. In 1970 Miller began a much more detailed examination of the site, digging 16 soil pits down to bedrock. In 1993 he added 21 new holes and then recorded soil depth and characteristics at all 37 pits. Far from the uniformity previously assumed, Miller found that the depth to bedrock across the study area varied from 1 to 14 feet, and soils ranged from coarse water-draining gravel to fine water-holding clay. Miller concluded that some of the superior growth recorded in the narrow spacings was a result of more favorable soil conditions.[4] Again, over time the forest revealed impacts of environmental variations that had not been predicted when the experiments began.

Later spacing experiments at Wind River attempted more control of variables. One such experiment was a study established in 1953 by George Staebler, who worked at Wind River before becoming director of forest research at Weyerhaeuser Company. Staebler chose a site at Planting Creek where trees had been planted in 1929 following a wildfire. He did a precommercial thinning of the 24-year-old trees at various densities and pruned back the crowns, attempting to find the optimum number of trees per acre to create a uniformly growing, easy-to-manage plantation. Reflecting the contemporary desire to grow perfectly uniform trees in perfectly uniform stands, Staebler concluded that "it would be desirable to have seedlings or saplings evenly spaced and far enough apart that every tree would make the maximum possible diameter growth for the site until the trees reach commercial size."[5] But trees resisted such intensive management. Reukema reported that in the first 10 or 15 years, the thinned trees in Staebler's study slowed their rate of height growth. Initially Reukema suggested that such thinning shock had been related to the exceptionally wide spacing of thinning on such a low-quality site. But 25 years after Staebler's thinning, the trees had increased their rate of height growth, especially at the wider spacings, and increased in diameter, once again confounding earlier conclusions.[6]

In the pursuit of ever-greater yields, researchers at Wind River questioned if thinning could be used in older trees to stimulate new growth and limit mortality from disease. Researchers had thinned 80-year-old stands at a low-quality site on the Panther Creek Division in 1938, removing trees that were dying from suppression, root rot, and insect attacks. By 1952 the stand had grown back beyond its original volume before thinning and was ready for commercial thinning. The second round of thinning salvaged trees that would have otherwise died and allowed more room for remaining trees to grow.[7] And although the sawlogs from this thinning were of relatively poor quality, lumber markets in the early 1950s were anxious for wood, and mills were already adjusting to smaller, lower-quality trees.[8] This study encouraged forest managers to make multiple commercial thinnings in older stands.

In addition, researchers tested thinning in a 110-year-old stand near Boundary Creek, where laminated root rot was taking a toll. They thinned plots heavily, lightly, or not at all. Thinning failed to slow the rate of mortality in the first decade of this study, but after three decades researchers found that the thinned stands maintained normal growth and the mortality rate had slowed.[9] The Boundary Creek thinning study prompted a sideline investigation when researchers saw evidence of earlier damage etched into the tree rings while examining stumps on some of the plots. Scientists examining the rings then found scars where bears had clawed into the sapwood when the trees had been 25–45 years old. So bears, like mice, were now considered pests in tree plantations and needed to be controlled.[10]

With increased interest in thinning, forest managers wanted help in choosing which trees would be removed and which would be left to grow. Staebler compared two sites where trees of different sizes and positions in the forest canopy had been removed. He found that in a densely growing, naturally seeded stand at Martha Creek, larger trees grew faster with more space thinned around them.[11] This was not surprising, since bigger trees have larger crowns and roots and thus greater growing potential. In a young plantation at Planting Creek, the changes in growth were not significant, perhaps because Planting Creek was a poorer-quality site and the trees there were more widely spaced to begin with. Revisiting the site 30 years later, Reukema again saw differences based more on site quality than thinning strategy and concluded that "failure to achieve positive response . . . may be an important result."[12] In 2001 Boris Zeide, a forestry professor at the University of Arkansas, concurred with Reukema's conclusion. Zeide reviewed 200 years of thinning research from around the world and found no evidence that thinning increased total volume. Instead, in most cases he found that thinning increases growth of individual trees at the expense of the volume growth of the entire stand. However, Zeide noted that trees in thinned stands tended to grow larger and produce better lumber while maintaining the aesthetics of an intact forest.[13]

Additional thinning experiments at Wind River examined whether trees that would otherwise die from competition could be salvaged and whether salvaging would increase the size of trees that were left, and thus increase profits. But the management goals of federal and private forests would diverge in the coming decades, and the purpose of thinning would diverge as well. By the end of the 20th century, thinning on industrial lands would be replaced by widely spaced plantations harvested at a much earlier age, and thinning in federal forests would be done to re-create diversity that had been lost in plantations.

But even toward the end of the century, there were still questions about thinning shock, questions that had nettled Wind River researchers ever since Staebler first described the phenomenon in the 1950s. When testing a 20-year-old plantation at Wind River in the 1970s, Miller and Reukema found that thinning reduced the rate of height growth by 25 percent, whereas fertilizing the thinned trees nearly doubled the growth rate. These and other studies prompted one of the most recent thinning studies conducted at Wind River. Researchers examining thinning shock and the role that nitrogen fertilizer played in overcoming it used a modern statistical design that allowed them to compare results with older studies. They chose as a study site a nine-year-old plantation planted with local seed near the stands where researchers had observed severe thinning shock and increased growth with fertilizer. They thinned plots to varying densities and added nitrogen fertilizer three years later. The researchers found that thinning shock on the new site was not as severe and response to fertilizer was not as dramatic as on the older study sites. They looked at a host of reasons to explain these differences, including age, site quality, amount of precipitation, and fire history. The variable that most consistently explained the difference was seed source.[14] The older studies were planted with seeds collected from outside the local area, which may have predisposed the trees to thinning shock and an exaggerated response to nitrogen fertilization. The younger stand had been planted with local seeds, reiterating decades of findings from the heredity studies that confirmed the value of using locally adapted seeds.[15]

The importance of locally adapted seeds was again demonstrated by Roy Silen's analyses of racial variation in Douglas fir and ponderosa pine. It was now generally possible to predict how fast and how big a tree would grow from a given seed source in a particular locality.[16] According to Silen, tree genetics were tightly adapted to their environment. A family of trees with clear superiority in one region would not be expected to flourish in another region. For example, trees native to the Cascades would not do well in the Coast Range, nor would trees native to the Coast Range do well in the Cascades. He argued that steady but slow improvement in tree growth could be gained based on the known performance of tree families that had been followed in the 50-year-old heredity study. Silen's extension of T. T. Munger's heredity study led him to propose a tree improvement program for the Douglas fir region based on long-term tests. He cautioned that such a program should not move too fast because he knew that early results could be reversed in later years. Eventually the program that Silen outlined produced over 2.7 million progeny from over 20,000 parent trees from breeding zones throughout Oregon and Washington. This was a huge accomplishment with two very important results: it produced improved seed that would grow Douglas fir trees at a faster rate, and it protected an enormous amount of genetic diversity. This genetic resource would be available for future experiments and to address yet unimagined questions. "Certainly, neither the plantations nor our concepts of forestry are going to remain the same in another 30 years," Silen wrote.[17]

Forest research was changing rapidly during the mid-20th century. The Society of American Foresters recognized the 25th anniversary of the McSweeney-McNary Act in 1955 with an examination of the future of federal forest research. The report predicted a fourfold increase in research funding in the coming quarter century, with new emphasis on wildlife, range, watershed, and recreation studies.[18] Moreover, federal forest research in the Pacific Northwest was shifting from field experiments to laboratory experiments. New forest science laboratories opened in Olympia and Wenatchee, Washington, and in Corvallis and Bend, Oregon. Staffed by technical experts, the labs were subdivided into specialized projects, such as high-elevation silviculture and watershed studies, and each was headed by a project leader in separate labs.[19]

Attention to long-term plots took a backseat to lab work. Reorganization had left Wind River with no researchers on-site; they were now working in labs off-site. The experimental forest was now administered through the new Olympia Forestry Sciences Lab under a project, headed by Miller, to study the silviculture of Douglas fir. Although ignored by most laboratory scientists, the Wind River Experimental Forest still held value for the silviculturists at the Olympia Lab. Since the early 1900s, scientists at Wind River had systematically studied the growth and yield of even-aged stands of pure Douglas fir in permanent sample plots they had established across the region. In 1963 Richard L. Williamson, a silviculturist at the Olympia lab, conducted a 50-year review of these permanent sample plots. He noted that when the plots were first established, "a full day by stage, row boat and foot travel was required to cover the 30 miles separating three plots on the Willamette drainage." Williamson concluded that the result of continued investment of time and effort was a detailed life history and an unexcelled view of typical stands occurring over millions of acres in the Pacific Northwest.[20]

But Williamson's assessment was not widely held among administrators. Just as stagecoaches and rowboats had given way to cars and bridges, field-based forest experiments were giving way to highly technical, laboratory-based research. In the era of Sputnik, field observation and permanent plots seemed to some people the quaint tools of a bygone era, no longer useful for faster-paced, larger-scale research. Computers sped the pace of testing by simulating growth and yield that would take decades to observe in nature. Many forest researchers rejected years' worth of data because the permanent plots did not follow modern statistical design (e.g., they were not randomized nor replicated, and some were too small). For many, the permanent plots were old-fashioned, out-of-date, expensive to maintain, and worthless to measure. Some recommended abandoning the plots altogether and pulling the tags that marked the carefully measured trees.[21]

In the mid-1960s, George Meagher, PNW director of timber management research, proposed reducing the size of the experimental forest by deleting the northern portion of the Trout Creek Division, including Trout Creek Hill, and liquidating its old-growth "decadent" timber.[22] Timber production was the ruling paradigm of the national forests at this time, and forest supervisors were looking everywhere for timber they could add to the harvest. Experimental forests had to prove their worth by providing income or knowledge that would help generate income. Otherwise, researchers risked being seen as squatters on valuable land, tying up forests that could be used to meet the nation's demand for timber. Six of the station's project leaders found no reason to object to Meagher's suggestion, as they had no plans for research on the Trout Creek Division. Only Silen countered with a research proposal. Silen agreed that the old-growth forest of Trout Creek Hill held no research potential in an era when regional timber production was increasingly focused on younger, second-growth forests. He also agreed that old growth was defective and should be cut.[23] But rather than simply cut the forest and decrease the size of the experimental forest, Silen proposed an experiment.

As a cone-shaped volcano, Trout Creek Hill could be delineated into study sites according to aspect and elevation. Silen outlined a harvest pattern that would create a series of age classes on Trout Creek Hill, which would provide a place for continuing research. Silen suggested that with little effort, cutting Trout Creek Hill in an experimental pattern would simultaneously relieve some of the pressure on the station to contribute to the harvest and provide opportunities for new research.

Silen recalled later that "Meagher may have been under pressure to propose a use for the area and increase cutting on the experimental forest."[24] Silen's plan would do both by contributing valuable timber to the national forest's account and by creating a greater array of age classes to study over a period of several decades. For whatever reason,

Meagher was intrigued with the idea, and in 1965 he proposed the idea to Ross Williams, supervisor at the Gifford Pinchot National Forest. As it turned out, Williams had been looking for an area on the Gifford Pinchot Forest that would provide opportunities to test a new piece of logging equipment. He proposed coordinating with the research station on a plan to remove old-growth timber from the experimental forest, and ended his proposal with "Time's a wasting and so is the timber."[25]

As the research station contemplated Williams's proposal, the screws tightened on the Wind River Experimental Forest. In April 1966, district surveys uncovered beetle-killed timber on the Panther Creek Division. Williams informed the station that it needed to salvage 100 million board feet of beetle-killed timber in the experimental forest.[26] The next year he requested cutting two units and building 2.3 miles of road in the Trout Creek Division.[27] Timber sales continued to chip away at the experimental forest, as efforts throughout the region were focused on liquidating old growth as quickly as possible. Old growth held no research value at the time. Private and public forest managers needed information about young second-growth forests, of which Wind River had few.

It was time to reconsider Silen's proposal. He and Miller drafted a plan for Trout Creek Hill. Unlike other incursions into the experimental forest that whittled away opportunistically at the old-growth forest, management on Trout Creek Hill could be carefully designed to create real opportunity for new research. They proposed a 60-year cutting plan to create a series of young forests that would create new sites for research at the experimental forest.

Silen outlined a need for research by asking how the Forest Service planned to convert its old-growth forests to high-yielding younger stands. How would forest managers address issues such as the conversion of brush and alder lands, natural regeneration, road standards, animal damage, and streamside slash cleanup?[28] Miller followed up with three options for Wind River: (1) maintain the status quo with little gained, (2) revive the experimental forest as a place to demonstrate new forest practices, or (3) give it back to the national forest.[29]

Miller had thrown down the gauntlet. It was time to consider the future of Wind River. Miller, Silen, Robert Tarrant, station director

Robert Buckman, and assistant director Robert Romancier retreated to the experimental forest and plotted its future. Plans for roads, harvest, and even rock quarries were being proposed for the site, and the scientists needed a clear research plan for the experimental forest or they would risk losing it to the national forest. To retain the experimental forest in its entirety, they agreed to increase research use and demonstrate its value for the future of intensive timber management with new experimental stands.

The first of these new timber management studies would be an examination of residues left behind from logging on Trout Creek Hill. Liquidating old growth was messy business. Only the logs had any market value, but clearcutting left behind broken limbs, defective big trees, and wasted undergrowth. Similarly, thinning created a disposal problem as culled trees piled up at the edge of thinned stands. These so-called residues were thought to impede regeneration, increase the threat of fire, and contribute to forest pests and diseases, and to many people they were eyesores.

PNW forest economist Thomas Adams outlined a study of residues in 1972. He saw value in residues as the demand grew for smaller material in pulp and paper manufacture and new wood fiber products. The energy crunch of the 1970s provided one more use for this seemingly wasted resource—as fuel for woodstoves warming the hearths of homes throughout the region. Better utilization of residues was important, he told an audience of foresters, because "we just have to make the land look better cared for."[30] Residues not only cluttered the land after thinning and harvest, but also choked streams and blocked the paths of migrating salmon. In the coming decades, the value of wood on the ground and in streams would be considered of ecological importance, but in the 1960s and 1970s, it was a leftover mess that needed to be cleaned up.

The first timber on Trout Creek Hill was sold in 1975, and cutting began soon after. The plan called for 480 acres to be harvested in 8 units, each with a different experimental residue treatment, to examine the effects on the survival and growth of planted seedlings. Researchers disposed of leftover residue by either broadcast burning, piling, and burning, or scraping the ground with a backhoe; they then followed their tests with different methods of planting. Further studies

Fig. 7.1. The curvilinear shape of clearcuts on Trout Creek Hill follows the contours of the shield volcano. (Courtesy of USDA Forest Service.)

examined contributions to the ecosystem if residues were left on-site for nutrient cycling, soil protection, seedling protection, and wildlife use (fig. 7.1).[31]

The Trout Creek Hill design created the possibility for more silvicultural experiments. By the late 1970s, many thousands of acres of clearcut forests throughout the Pacific Northwest, public and private, were being planted with monocultures of Douglas fir at 10-foot spacings. It was a practice based mostly on habit and very little research. Dean DeBell, a silviculturist at the Olympia Lab who would later take over management of Wind River, recognized that economic and social changes in the region would soon require new knowledge about using native species other than Douglas fir in pure and mixed plantings and at various spacings. He saw an opportunity to dovetail new silvicultural experiments with the residues study at Trout Creek Hill.[32]

This study was unusually large in scope and would not have been possible without the help of the Wind River Ranger District in establishing the new experimental plantations. DeBell tested spacings and species that might be used in future management regimes: Douglas fir, western hemlock, western red cedar, western white pine, and noble fir in various mixed plantings, including Douglas fir interplanted with alder, which followed Tarrant's work at Planting Creek. All the seed was from local sources. DeBell wanted to know how growth in mixed species stands would compare with pure stands, a research approach very different from previous silviculture research, which had focused on single-age, single-species stands almost exclusively. In the 19th century, mixed species planting had been advocated by Bernhard Fernow, who had argued that they offered greater security against damages by wind, fire, frost, snow, and diseases, and yielded a larger amount of wood compared to pure plantations,[33] but the practice was never pursued in the Douglas fir region. DeBell understood the biological and economic risks of growing trees in a monoculture, and he foresaw the forest economy diversifying in the Douglas fir region, making room for other kinds of timber.[34]

The big Trout Creek Hill experiments were not the only research projects the Wind River scientists outlined for the experimental forest. No project better fit the can-do attitude of science and technology than the idea to haul old-growth timber from the forest with the use of military helicopters and balloons. Known as the Heli-Stat, the contraption was jerry-built from a Navy surplus balloon lifted by four helicopters. Promoters won congressional funding with the promise that with no roads to build in the forest, huge savings would be made. Congress took the bait and funded the project in 1980. Wind River would be the test site.

In order to carry logs aloft, scientists needed to determine just how much these old-growth trees really weighed. In an elaborate experiment at Wind River, branches of 32 Douglas firs and 29 western hemlocks were carefully removed from standing trees and the limbs were lowered to the ground to be weighed. Adding limb and crown weights to known log weights, the scientists were able to calculate the weight of potential payloads.[36] But the Heli-Stat was never tested. Just before deploying to the West Coast, the Heli-Stat crashed, killing one of the helicopter pilots during a test flight in Lakehurst, New Jersey, the same airfield where the Hindenburg dirigible had crashed 40 years earlier (fig. 7.2).[37]

It may have seemed to some that Wind River's future was headed for a crash. Nonetheless, visiting researchers still flocked to the experimental forest to learn about Douglas fir, which was widely planted throughout the world's temperate zones by the 1960s. Wind River had become a showplace for foreign visitors interested in seeing

Fig. 7.2. The Heli-Stat being lifted into the air by four helicopters in Lakehurst, New Jersey, date unknown. (Courtesy of Piasecki Aircraft.)

the original site of classic thinning and spacing studies. Forest managers came from as far away as Europe and New Zealand to learn how to grow Douglas fir, one of the best timber trees in the world. But just as the old experiments were beginning to reveal unexpected results, the hidden costs of intensive forestry were beginning to appear in the Northwest forests.

Chapter Eight

Forest Science in Transition

❧

It is imperative that the research balance of the future include liberal attention to values of the forest in addition to timber production.

—Charles Connaughton[1]

The prosperity that followed World War II brought a new public interest in natural resources, more for their aesthetic and recreational value than their commodity value. According to historian Samuel Hays, this new perspective was closely associated with rising standards of living, levels of education, and access to leisure time to complement urban life. The environmental awareness that was emerging in the 1960s and 1970s saw the forest as an extension of the human experience, a vision that would work its way up through grassroots organizations to gain increasing political influence.[2] It was a vision of untrammeled nature, celebrated in calendar art and coffee-table books that portrayed the forests of the Douglas fir region as cathedrals of moss-draped old-growth trees. The sight of clearcuts clashed with that vision and interfered with the human experience of the forest. Scientists would later study how clearcuts changed the way forests functioned, not just the way forests looked, but at the beginning of the environmental movement, it was aesthetic values that were most out of tune with intensive timber management and the practice of clearcutting.

By the 1970s, even-aged management had become the standard practice throughout most of the Douglas fir region. The prescription was uncomplicated: clearcut, slash burn, and replant with proven stock in single-age, single-species plantations. Throughout the process, the trees were protected against pests and diseases by chemical pesticides and herbicides. Not only did this prescription promise to yield ever larger volumes of wood, but some of its boosters promised multiple benefits from the improved forest, including open areas for deer to browse and cleared vistas for visitors who could now access the national forests by way of thousands of miles of logging roads.[3]

In 1972 Oregon State University's (OSU) College of Forestry hosted a symposium on even-age management of forests as part of the Second North American Forest Biology Conference. Two forest scientists were invited to speak about the effects of harvest methods on regenerating even-age stands: Jerry Franklin, a forest ecologist at the PNW Research Station in Corvallis, and Dean DeBell, who at the time was a silviculturist with Crown Zellerbach Corporation. Franklin and DeBell, the only nonacademic forest researchers to make presentations at the conference, had been asked to collaborate on a paper, with Franklin to focus on western forests and DeBell on southern and eastern forests, where he had worked extensively in the hardwood bottomlands (figs. 8.1, 8.2).

Fig. 8.1. Jerry F. Franklin, 1991. (Courtesy of Jim Harrison, Heinz Foundation.)

*Fig. 8.2. Dean S. DeBell, 1985.
(Courtesy of USDA Forest Service.)*

The two researchers worked together, reviewing the literature and reflecting on their different field experiences. Both Franklin and DeBell found that although there may be *economic* reasons to clearcut large sections of the forest, there were no *ecological* reasons. They concluded that economic and social considerations, rather than ecological considerations, determined the choice of cutting. "There is no ecological necessity for large patch or continuous clearcuttings to regenerate most types, species, and sites including Douglas fir," they concluded.[4] But this conclusion flew in the face of the current belief that clearcutting was required for the successful regeneration of Douglas fir. With protests over clearcutting growing around the country, Franklin and DeBell's paper found boosters on both sides in an increasingly polarized debate. It hit a raw nerve in parts of the timber industry and the Forest Service, although George Staebler, now at Weyerhaeuser, and Clarence Richen at Crown Zellerbach praised the paper.[5] National newspapers picked up the story, and soon reporters from across the country were calling experts to weigh in on the researchers' conclusions. Franklin recalled that the *Wall Street Journal* quoted William Hagenstein, the director of the Industrial Forestry Association, who dismissed the paper as simplistic and its authors as "young scientists without experience in forestry."[6] In the coming years, DeBell and Franklin would gain plenty of experience.

At about the same time as the conference in Corvallis, a few turkey hunters in West Virginia were protesting clearcutting in the Monongahela National Forest. Claiming that clearcuts violated the very foundation of the Forest Service, the Organic Act of 1897, the turkey hunters took their protests to court. Similar lawsuits followed, threatening to reduce national forest timber harvests by at least half.[7] Clearly what was needed was a legislative solution, which came in the form of the National Forest Management Act of 1976. The act allowed clearcutting in national forests but mandated greater public involvement in Forest Service planning. This was one of a suite of environmental legislation that would change federal forest research and management in the years to come.

DeBell and Franklin characterized a new generation of forest researchers whose work would be intertwined with social and political changes in the Douglas fir region. DeBell had grown up in southern New Jersey, where he enjoyed fishing, hunting, and camping. "My dad suggested forestry as a possible career," he remembered. DeBell pursued the field at Duke University's School of Forestry. Like T. T. Munger and Leo Isaac before him, DeBell headed west following graduation, where he worked on the Wind River District with the timber management crew. He and his wife moved into a small cabin on the banks of Wind River, where they began a deep and lifelong connection to the river and the experimental forest. "One winter day our crew visited Crown Zellerbach's pulp and paper mill at Camas and a chance meeting with CZ's forest research supervisor stimulated my interest in forestry research," DeBell recalled. This interest led him to research in silviculture and applied forest ecology with Crown Zellerbach, the Forest Service's Southeastern Forest Research Station, and eventually to PNW Research Station in 1975.[8]

Jerry Franklin also had a deep connection to Wind River, stemming from his childhood memories growing up in the nearby paper-mill town of Camas, Washington. "I played and scouted on the edge of my home town in forests that had regenerated following the Yacolt Burn," Franklin remembered. He became aware of forestry during family camping trips, particularly during a weeklong stay at Government Mineral Springs campground in the upper Wind River valley. "By the time I was ten I had decided that a forester's life was the one for me," Franklin said. While a student at OSU's College of Forestry, Franklin found his specialty in research as a student trainee at the PNW Research Station. But as a forest researcher in the 1960s, there was no reason to study old-growth forests. The future, it seemed, was in even-aged management of plantations. "We grew up thinking of old forests as biological deserts or cellulose cemeteries," Franklin recalled. "We climbed over huge piles of downed logs and woody debris, and we didn't think about anything other than how to get rid of it, how to liquidate it."[9]

By the late 1960s, however, Franklin began to pay attention to the messy old forest. In 1969 the National Science Foundation granted funds to study coniferous forests in the Pacific Northwest as part of the new International Biological Program. The funds were to be shared between University of Washington in Seattle and OSU in Corvallis. The Corvallis group included forest scientists from OSU's College of Forestry and the PNW Research Station. Some in the group wanted the funds to go toward studying the ecology of younger, plantation-style forests—the forests of the future. But others, including Franklin, pressed for a cooperative study of old-growth forest ecosystems "before they disappeared."[10]

The Corvallis group centered its old-growth research at the H.J. Andrews Experimental Forest in the Oregon Cascade Mountains, about a two-hour drive southeast of the OSU campus in Corvallis. Attracting funds from the National Science Foundation and elsewhere, the research program thrived with a strong partnership among federal and university scientists. The university partners helped leverage funds, broadened funding sources, and provided a rich interaction between applied and basic research. In turn, the federal partners offered a vast land base and a place to test theories in the field. In contrast, without research and funding partnerships, the Wind River Experimental Forest began to become almost a backwater of forest research, a museum of old experiments. In retrospect, it is tempting to imagine the research that might have been developed at Wind River, such as experiments comparing clearcutting with other harvest methods. But in the 1960s and 1970s, clearcutting was the established paradigm, a given assumption, not a variable to be tested. Research on harvest methods was largely restricted to examining aspects of clearcutting. Research of alternatives, such as the selection cutting that Burt Kirkland and Axel Brandstrom had proposed a generation earlier, would be years in the future.

However, forest research, perhaps prompted by growing public interest in the environment, was beginning to examine the connection between forest practices and environmental quality. The study of ecology had gained scientific momentum since its early studies of landscape succession in the 1930s. In the early 1970s a consortium of 75 colleges and universities formed the Institute of Ecology and proposed that Wind River be a part of a network of experimental ecological reserves. The institute's purpose was to enhance ecosystem studies and secure sites for long-term research "on the structure and function of natural ecosystems and on the effect of man's perturbations on these systems."[11] PNW Research Station director Robert Tarrant agreed to the proposal. But with no university connections and no access to federal grants, Wind River remained isolated from most mainstream ecosystem research, and little came of its involvement in the institute.

Beyond Wind River, forest research was opening in new directions, fueled by new and powerful computing technology. The 1950s and 1960s had taken science into the laboratory and up into space. The next two decades brought forest science outdoors once again, on a large scale and with tools that could extend the reach of science across landscapes. Geographic information systems enabled scientists to model forest processes on a larger scale and with more complexity than ever before. Along with improved instruments, statistical design matured with new emphasis on the importance of randomizing samples and replicating plots. Computer models made it possible to fast-forward the possible long-term effects that intensive forest management could have on future forest productivity.

New tools helped reveal ecological changes occurring in the Douglas fir region that challenged the technological ideal of a fully regulated and predictable forest. *Biodiversity* and *complexity* became the new watchwords in forest science. Although studies on clearcutting and regeneration methods continued, more attention became focused on the visual and ecological impacts of timber harvesting. Forest scientists began questioning the effects and the ethics of their work.

Public involvement, social change, public controversy, and population pressure were all new topics debated at professional meetings and in the pages of professional journals, such as the *Journal of Forestry*.[12] "Forest research has done well enough with technology but rather less well with ecology and awareness of all the social values of the forest," one author wrote. "Applied research for the national forests should not take precedence over long-range basic ecological research, as has so often happened in the past."[13]

But little new research was making its way to Wind River in the early 1970s. The National Forest Management Act required a formal process of planning and attention to multiple aspects of the forest, including research. Although the development of forest plans proceeded slowly, there was nothing slow about the rate of harvest in the surrounding national forest. Harvest in the Gifford Pinchot National Forest was 464 million board feet in 1975,[14] and the expectation was to keep up this level of harvest into the future. Salvage cuts exposed the edge of old-growth stands on Trout Creek Hill, making them vulnerable to blowdown, and aerial fertilizer drifted over research plots. The nursery cleared land up to the edge of the research natural area, exposing the area to drifting herbicides. Franklin became alarmed by the increasing pressure on the experimental forest. He wrote to DeBell and to assistant station director Edward Clarke, emphasizing the need to develop explicit objectives for management and use of the Wind River Experimental Forest. "Given the difficulties of obtaining research areas now and in the future we can't let either Division go back to ordinary national forest land," Franklin wrote. "We need a plan."[15] Despite the convincing demonstration of research value at Trout Creek Hill, Franklin and others still feared that the national forest could make a convincing proposal to clearcut other parts of the experimental forest if there was no visible research use there.

In July 1979 Robert Buckman, now the Forest Service's chief of research in Washington, D.C., asked all research station directors to consider the full range of options within experimental forests, including timber harvesting. "We want to be careful to not impact on the integrity of our experimental forests for the intent for which they were established," Buckman asserted. "However, at the same time, we do want to look at opportunities to provide timber from all

Forest Service lands in a positive manner."[16] In response to Buckman's directive, Tarrant called for a review of the Wind River Experimental Forest in August 1979 and invited Roy Silen, DeBell, and Franklin to participate.

The struggle to prove Wind River's worth would be measured in research volume versus timber volume. For the review, Franklin outlined in detail all the research he and others at OSU and the PNW Research Station lab in Corvallis were conducting at the Wind River Experimental Forest. Most of what he outlined were studies in coordination with the ecosystem research at the H. J. Andrews Experimental Forest, ranging from forest succession to woody debris dynamics and decomposition. These studies required areas of forest that would be protected for research; in short, they required experimental forests. In addition, Franklin listed the silvicultural research at Wind River, including Silen's continued work on heredity, ponderosa pine, and arboretum studies. Some cutting might be necessary to create new stands for research, but Franklin emphasized the need to hold a "substantial chunk of the remaining old-growth . . . as an experimental reserve . . . for future manipulative research."[17] In his final report, Tarrant noted that the timber harvest and associated silvicultural studies that DeBell and others were beginning on Trout Creek Hill were "in line with . . . Buckman's July 26, 1979 memo to increase harvest on experimental forests where appropriate."[18]

However, the fate of the Panther Creek Division was still up in the air. In June 1980 DeBell convened a "blue-ribbon" interdisciplinary committee to develop a comprehensive management plan for the entire experimental forest. Despite the pressure the researchers must have felt, they saw this as an opportunity to try new approaches in forest research. DeBell's committee included experts in soils, geology, flora, insects, and fire—in short, every aspect imaginable in the forest ecosystem. Such an interdisciplinary approach, although it would become a common practice years later as ecosystem management, was a very different way of doing business at that point in time. Despite the researchers' optimism, a broad-based management plan for Wind River remained little more than an idea. The planning proceeded slowly. Involvement of so many expert collaborators proved difficult, as they were pulled away for many other assignments. In the end, a draft plan emerged from the concentrated effort of DeBell, Franklin, and Richard Woodfin, who managed the residues program at the PNW Research Station in Portland and oversaw operations at the Wind River Experimental Forest during this time.

Meanwhile, formal planning processes were grinding away in other places within the national forests and across Region Six. Old-growth forests were a planning element that seemed to have everyone stumped. Old-growth forests, whether they were seen as cathedrals or graveyards, were more clearly defined in the public imagination than they were in science. What were the defining characteristics of old-growth forests?

Franklin, in collaboration with Richard Waring, an OSU forest science professor, had described the features of the massive evergreen forests of the Pacific Northwest and had characterized the old, native forests of the Douglas fir region. They reported that these old forests were unique among temperate forests of the world, distinctly adapted to the Northwest climate and geography, and dominated by a few widely distributed conifers that could grow very large and very old. The trees' age and size buffered them against stresses of drought and fire.[19]

Managers from western Oregon's Siuslaw National Forest, the nation's most productive forest for growing timber, were among the first to recognize that old growth would be a controversial forest planning issue. Forest planners needed a clear scientific definition of the forest type that was sometimes called cathedral, virgin, ancient, or simply old growth. They called on Franklin for help.

Franklin convened a group of researchers at Wind River in 1977 to hash out a definition of old-growth forests. He chose Wind River purposefully. It was a place with a rich research history, an old and familiar place to develop new ideas, with a long record of experimentation on the very forests in question. It was a place where the scientists could do more than examine tables of data. They walked among the old trees, kicked the dirt, and talked late into the night in the old training center that the Civilian Conservation Corps (CCC) had built more than 40 years earlier. After a few days of deliberation, the scientists began to describe what they saw as the ecological characteristics of old-growth Douglas fir forests. Eventually a definition emerged, characterized by old age and immense size, layers of crowns, and a forest floor littered with woody debris. What was stunning about this definition is what it was not: an estimation of board feet of lumber. "Viewed objectively, old growth forests are neither the paragon of virtue and beauty imagined by some, nor the purposeless wastelands imagined by others," Franklin said. "We sometimes emphasize only attributes consistent with our current management goals or policies . . . We need to . . . incorporate attributes of natural forests in our schemes for intensive management."[20]

The Wind River meeting may have been what Aldo Leopold had meant by a stirring of an ecological conscience.[21] The studies that followed this meeting more than defined old-growth forest types; they defined ecosystem studies for the next two decades or more. In particular, new studies began to explore the immense size and longevity of species, biomass accumulations, and the overall importance of coarse woody debris in the giant, old forests of the Pacific Northwest. The definition—and the phenomenon—of old-growth forests included a parade of plants and animals that had been overlooked or discounted when the forest was valued primarily as a timber factory. It wasn't long before old-growth studies expanded to examine characteristics of natural forests of all ages. Some of those studies coalesced into a 1991 publication that examined the vegetation and wildlife in unmanaged young and old Douglas fir forests. The publication's editors, led by PNW wildlife scientist Leonard Ruggerio, described the crossroads faced by forest managers: "Should ancient forests be protected for their aesthetic appeal and because they provide a broad range of ecological values, including the most amenable environment for some plants and animals? Or, should they be harvested because the revenue they provide affects the economic stability of the entire region? These questions encapsulate one of the most heated and socially significant conservation and natural resource management debates of this century."[22]

The Ruggerio publication included examinations of wildlife habitat, fire history, and forest fragmentation in natural stands of Douglas fir. The studies required representation of habitats within all ages of natural forests in parcels large enough to be statistically meaningful. Whereas silviculture studies had used plots on the scale of fractions of acres, wildlife studies needed plots on the scale of miles. The researchers had a problem finding miles of forest that represented all forest ages. They found plenty of second growth in plantations and some protected pockets of old growth within the research natural areas, but it was surprisingly difficult to locate young natural forest stands within the region. After a thorough search of the region, they found what they needed at Wind River in the reburned, naturally reseeded areas of the Yacolt Burn.

As wildlife habitat research and old-growth ecosystem studies renewed interest in Wind River, DeBell's blue-ribbon team finally finished the Wind River Experimental Forest Management Plan in 1987. The document was impressive for its efforts in promoting interdisciplinary work and its compromise between forest manipulation and preservation. The plan called for a balance between creating new stands and preserving existing stands, and it put new emphasis on biological diversity. It suggested a two-pronged approach to research at Wind River, one focused on silvicultural approaches to creating new stands of mixed species, the other on understanding the ecosystem processes of existing natural stands.[23] Much of the research described in the management plan was oriented toward harvesting, regenerating, and growing new stands of timber, the regime that existed on most industry and national forestland at the time. The plan included a study of how roads affect soil erosion and watershed function, and how insects affect seed production, regeneration, and growth. In terms of recreation, the plan centered mostly on aesthetics and visitors' reactions to the visual impact of clearcuts. But among the studies of management-oriented problems were also questions of basic research.[24] Potential new research topics included a frontier almost unexplored at the time: the forest canopy and its interactions with the atmosphere.[25]

Although the Wind River management plan saw the issues of conversion and preservation as complementary, the rest of the world saw them as polar opposites. Events beyond the borders of Wind River boiled into conflicts across the Douglas fir region. Landslides linked to road building in steep forested terrain slumped tons of sediment into salmon streams in western Oregon. Chemical herbicides used to clear brush out of timber plantations poisoned people living in adjacent rural communities. Logs sold at below cost from national forests were shipped overseas to be manufactured. Despite the collaborations of scientists at Wind River, the two approaches to forest science—silviculture and ecosystem studies—began to diverge in the minds of others. Entrenched camps adopted one scientific approach and rejected the other. Such polarization would drastically change the direction of forest research in the Douglas fir region during the next few years. By the time the Wind River management plan was finally completed, the political landscape had changed enough that much of the plan could not be implemented.

Munger did not live long enough to see the new management plan finished. Early on, he had understood that there was something to be learned by studying the old-growth forest. It was a driving reason for him to establish the research natural area in a 450-year-old stand of Douglas fir and western hemlock within the experimental forest. When he died in 1975, the research natural area at Wind River was renamed the Thornton T. Munger Research Natural Area.

In 1983 Franklin and DeBell—the ecosystem scientist and the silviculturist—measured the old forest together, and they saw different lessons in the data they collected. Franklin's paper, which DeBell coauthored, emphasized the ecosystem processes and functions within the forest—the change in population, mortality rates, and stand development. He calculated that the slow succession of the forest would take 850 years for the Douglas fir to die out and leave a "climax" western hemlock stand in its place. DeBell's paper, which Franklin coauthored, emphasized changes in stand characteristics as the forest grew older, describing the loss of Douglas fir, the emerging dominance of western hemlock, and increasing numbers of Pacific silver fir in the lower canopy. Both scientists recognized that the old-growth forest was dynamic, not a static fossil of ancient trees. But as far as wood production was concerned, both recognized that the old-growth forest's greatest growth potential had passed.[26]

Equally important to their findings was the collaboration of the two scientists. The two papers were intended for different audiences—Franklin-DeBell addressed a scientific audience and was published in the *Canadian Journal of Forest Research*; and DeBell-Franklin addressed foresters and managers and was published in the *Western Journal of*

Applied Forestry. Franklin and DeBell, who would come to represent two different approaches to forest science, began as collaborators in the old-growth forest at Wind River. They looked at the same forest, measured the same attributes, and recognized two different approaches to studying that forest. To these scientists, silviculture and ecosystem studies were two complementary lenses with which to see the forest more completely.

Chapter Nine

New Plans and New Restrictions

> *The most serious threat to the spotted owl in Oregon is the gradual elimination of its preferred habitat (old-growth and mature forests).*
>
> —Eric Forsman[1]

Despite collaborations at Wind River, forest science in other parts of the Douglas fir region continued to bifurcate. Differences in scientific approach between silviculture and ecosystem studies were made deeper and wider by a growing battle that swirled around the remaining stands of old-growth forest in the Douglas fir region. The flash point of the battle was the northern spotted owl (fig. 9.1).

The work of one scientist flew into particular prominence. Eric Forsman, a graduate student at Oregon State University (OSU) in the 1970s and later a PNW wildlife scientist, documented the relationship between this little-known bird and its dwindling habitat in old-growth Douglas fir forests.[2] Evidence eventually landed the northern spotted owl on a long list of imperiled species to be considered for protection under the federal Endangered Species Act. Such protection threatened to put large sections of the remaining old-growth forest off-limits to logging. Communities dependent on timber from federal forests would no longer be able to cut old-growth stands where owls were found. Mills would close and people would lose their jobs. On June 25, 1990, the northern spotted owl landed on the cover of *Time* magazine with a story headlined "Owl vs Man."[3] The owl had become a poster child for more intractable problems of economic and ecological diversity in the Douglas fir region. As such, the bird symbolized a much larger debate over the objectives of federal forest

Fig. 9.1. A young northern spotted owl. (Courtesy of Eric Forsman.)

management. Fueling these so-called timber wars, rancorous publicity polarized Northwest communities by posing the debate as owls versus jobs, rather than the mutual dependence of both owls and jobs on the long-term sustainability of forests. The concept of dependence was debated, as was the definition of sustainability. Previously, the concepts of dependence and sustainability of wildlife habitat had been considered primarily scientific ideas. Now they had great social and political importance. The region's popular identity began a subtle shift from the "Douglas fir region" to the "spotted owl region."

Up until this time, much of forest science at Wind River had focused on how to grow trees quickly and efficiently for maximum timber harvest, especially as Congress ratcheted up harvest levels. Maximizing wood production did not require attention to wildlife, aesthetics, or biological diversity; it required attention primarily to what was *removed*, not what was left behind.

When the objective had been simply to maximize wood production, silviculture research had succeeded in providing efficient methods to grow more timber faster. One prescription seemed to fit most timber harvest conditions: clearcut, burn, and replant with a monoculture of Douglas fir. With that success, research had become narrowly focused on clearcutting and even-age management that had been proposed in the 1930s. T. T. Munger and Leo Isaac had marginalized alternative harvest strategies such as selective cutting and uneven-age management. In addition, none of the natural regeneration methods had proved to be as efficient as planting nursery-grown seedlings, especially with genetically improved stock. Planting blocks of trees all the same age reduced the need for individual assessment of trees; stand-level management decisions could be made and clearcutting provided an economical way to establish quick-growing, even-aged stands that could be cut in increasingly shortened rotations. It was a system that emphasized efficiency and uniformity.

Because the process of clearcutting, burning, and planting was simple and effective and minimized costs, it was applied almost everywhere. "It was a short step to the belief that this was the only way to do things," wrote Robert Curtis and Andrew Carey, a research biologist at the PNW Research Station lab in Olympia. Looking back, they concluded that this "clearcutting dogma . . . has been a political and social disaster."[4]

Throughout the Douglas fir region, federal forest managers struggled with the problems and long-term risks associated with clearcutting. Logging roads carved into steep hillsides could slump into streams and bury fish habitat. Replanting trees after harvest was difficult and sometimes unsuccessful on high-elevation and marginal sites, leaving clearcuts bare and vulnerable to weeds. Many people protested the sight of clearcuts and the use of aerial chemical sprays. Mill closures in Oregon and Washington fanned the flames of controversy over how federal forests were being managed, as the rate of so-called inventory depletion threatened the long-term sustainability of the forest industry in the Douglas fir region.[5] Public debates pitted long-established commodity producers against those who sought quality of life in natural settings, and each side claimed its base of scientific findings. Economic concerns met head-on with environmental and social concerns reflected in new environmental legislation. New objectives for federal land management were turning forest research upside down.

The 1960s and 1970s had seen newly emerging environmental values codified in a series of landmark legislation, including the Clean Air Acts of 1963 and 1967, the Wilderness Act of 1964, and the Land and Water Conservation Fund Act of 1965. The National Environmental Policy Act of 1969 required federal agencies to evaluate a range of alternatives when proposing actions that could have a significant impact on the environment. The Endangered Species Act (ESA) of 1973 provided statutory protection to plant and animal species threatened with extinction, and it became a powerful tool that mandated that consideration be given to protecting endangered species and, by extension, to biodiversity. In addition to the ESA, the National Forest Management Act of 1976 put additional emphasis on managing forests for biological diversity, wilderness, aesthetics, and wildlife.

The interdisciplinary pressures of forest planning brought a considerable number of professionals into the Forest Service with skills beyond traditional timber production. The Forest Service began to attract into its ranks a new generation of expertise, including anthropologists, economists, botanists, and wildlife biologists, to help prepare the plans and impact statements required by the new laws. The Forest Service had long had a strong esprit de corps built on shared values, experience, and training. New perspectives from people with diverse backgrounds and training began to reshape the culture of the Forest Service, a culture that had managed to stay largely unchanged since its inception under Gifford Pinchot. According to Samuel Hays, as forest planning proceeded, new questions of environmental integrity and new standards of scientific assuredness appeared.[6]

In earlier decades, when timber production had been the primary purpose of forest research, wildlife scientists focused on animal pests, such as the bears and mice of earlier studies, and promoted wildlife such as deer and elk for their value as harvestable resources. New legislation increased the scope of forest wildlife management beyond the study of trout, pheasants, and deer to include nongame animals such as amphibians, mollusks, and songbirds. According to Jack Ward Thomas, Forest Service Chief from 1993 to 1996, the net effect of many new environmental laws was to make protection of biodiversity the *de facto* overriding goal of the national forests.[7] The emphasis of federal forest management was beginning to switch from the timber that was removed from the forest to the plants, animals, and landscapes that were left behind. New silvicultural prescriptions were needed to meet new objectives of biological diversity, but there was no existing body of research from which to develop new prescriptions.[8]

Up to this point, much of traditional forest science had focused on experiments on small tracts of land and yielded results that could be applied in controlled applications in managed stands. New large-scale modeling technology had extended scientific thinking to larger spaces and time spans. With this more extensive view came more uncertainty that forests could be controlled in a predictable manner.[9] Whereas *removing* trees and residue had been the focus of earlier research, now the emphasis was on *retaining* trees, residue, snags, and downed wood that would contribute to nutrient cycling, soil stabilization, and wildlife habitat. Living trees were retained as a legacy, not only for regenerating new trees but also for rebuilding wildlife habitat and maintaining forested spaces for people (fig. 9.2). The old-growth forest, once dismissed as unproductive, was now central to new objectives emphasizing biodiversity, ecosystem processes, and aesthetics.[10]

Silvicultural research had lost time and potential knowledge by pursuing single-age, single-species management single-mindedly. For much of the 20th century, silviculture had been used to minimize diversity, not enhance it. As biological diversity became increasingly important, federal forest managers found that there were few studies that could help them reestablish forests with trees of various ages, sizes, and species that would be habitat for various kinds of wildlife. As forest

Fig. 9.2. Green tree retention on a thinning study site at the Panther Creek Division, 1987. (Courtesy of USDA Forest Service.)

management priorities shifted from timber production to ecosystem protection, some forest researchers in the middle of their careers saw a complete reversal of the objectives that guided their work. Along with new forest values, new concepts such as environmental ethics, ecosystem management, and adaptive management slipped onto the radar screen of forest science. These new concepts had roots that went back as far as the older work of Pinchot, Munger, and Isaac. Pinchot had contributed an ethic of conservation and the concept of sustainability; Munger had widened forest research to include ecosystem elements and landscape scales; and Isaac had developed a careful approach of testing and retesting and admonished against one-size-fits-all forest management. These concepts, marginalized for a while during an era of intensive timber management, reemerged in the 1990s.

Whereas scientific gatherings had once reported results from thinning experiments, now there were plenary sessions with discussions of forest legacies and social responsibility. One of the organizing concepts that coalesced around these new topics was New Forestry.[11] Championed by forest ecologist Jerry Franklin, who had joined the University of Washington faculty in 1986, New Forestry was an effort to use silvicultural methods to create forest stands with structural and functional complexity. Instead of embracing a single blanket solution for all forest management, New Forestry would design site-specific prescriptions for a wide variety of landscapes and objectives, looking beyond individual stands to consider diverse patterns across the landscape and through time.

As promising as they sounded, the principles of New Forestry were a big leap from scientific theory to management application, a leap that would require a foundation of field experimentation. In response to that need, Dean DeBell and Curtis outlined a possible role for silvicultural research in New Forestry in 1993. They critically examined forest research and management that pursued maximized timber production with reliance on clearcutting, ever shorter rotations, the adoption of forest practices from low-elevation to high-elevation forests, and the dominance of single-species, single-age plantations—in other words, the status-quo forest management for much of the Douglas fir region during the second half of the 20th century. DeBell and Curtis argued that longer rotations, diverse planting, and a variety of harvest regimes could provide more options and varied outcomes that would satisfy new expectations for managing the federal forests. They also argued that silviculture could provide solid understanding of forest processes at the stand level, and stand level dynamics were essential to understanding larger landscapes.[12]

"What has been loud and clear in the writings through the last century is that there isn't a one-size-fits-all harvest regime for the Douglas fir region, and yet on the ground that's what we've been doing," DeBell wrote. "The 'toolbox' for meeting our needs out of the forest is much larger than we think."[13] For decades, the toolbox of silviculture had been applied to one job: the creation of even-age Douglas fir stands managed for timber production. Much of the silvicultural research at Wind River had helped forest managers do that job efficiently, and there was little interest in applying the tools and techniques of silviculture to anything else. That disinterest changed when the overarching goal of management in federal forests changed from timber production to biological diversity.

While forest science in the Douglas-fir region was changing course, harvest levels continued to increase. With policies and goals that seemed to contradict one another, armies of private consultants, fund-raisers, and political activists were poised at opposite extremes of what Thomas later described as a new "conflict industry." Thomas was in a unique position to assess the savagery of the conflict. As a PNW wildlife scientist and soon-to-be chief of the Forest Service, he had been enlisted by Congress to oversee the drafting of a plan to protect the northern spotted owl. This initial plan led to an escalating series of reports and lawsuits charging that the Forest Service was violating environmental laws.[14] Lawsuits led to scientific studies, reports, plans, more lawsuits, and the eventual rewriting of forest management in the Pacific Northwest.[15]

Images of destitute timber communities on one side and smoldering clearcuts on the other paralyzed decision makers. Congress stepped in and again asked Thomas to head a scientific committee to develop management alternatives that could break the deadlock. The Scientific Panel on Late-Successional Forest Ecosystems, better known as the Gang of Four, was made up of K. Norman Johnson from OSU's College of Forestry, John Gordon of Yale University's School of

Forestry and Environmental Studies, Franklin, and Thomas. The group was summoned in 1991 by two committees from the U.S. House of Representatives to report on the extent and character of late-successional and old-growth forests, including wildlife and fish, on federal lands within the range of the northern spotted owl.[16] The panel delivered its report to Congress, but again, decision makers were paralyzed. Scientific studies stacked up higher and higher. Eventually, there were more than a dozen lawsuits pending or in court involving the northern spotted owl. The area of study expanded with every report, and forest science also expanded to include the full scope of the Douglas fir region, eventually encompassing all the natural and human communities within the region.[17]

The spotted owl conflict eventually halted all federal timber harvests in the region and got the attention of the president of the United States. In 1994 President Clinton and Vice President Gore visited with the region's forest scientists and representatives from the forest industry, timber communities, and environmental groups.[18] By presidential decree, the Forest Ecosystem Management Assessment Team (FEMAT), headed by Thomas, undertook the task of drafting a plan to save both endangered species and a failing timber industry. The Northwest Forest Plan that grew out of FEMAT would have profound implications for forest research as well as management.[19]

The process of drafting a plan brought the role of science and scientists under public scrutiny as never before. Forest scientists were no longer debating within their own circle of professionals; their debates were public, and their work was dragged into the political spotlight for review. The value of the work of researchers who had spent their careers successfully meeting the objectives of the old paradigm was suddenly called into question. Many researchers could not help but take these criticisms personally. No matter how successfully the old approach to forest research had met the old objectives, it was not enough to meet the new ones. The new laws put new demands on forest scientists to provide information that previously had not been part of their research agenda. Highly polarized conflicts within society began to claim whatever available science could support their cause; scientists were caught in the middle.

To meet President Clinton's direction for a plan that would protect jobs, provide timber, and protect the habitat of the spotted owl and the marbled murrelet, the Northwest Forest Plan designated areas of federal forest as the matrix where timber cutting was allowed, and reserved land for protecting riparian areas and species of concern. Consequently the entire Wind River Experimental Forest was designated as a Late-Successional Reserve designed to serve as habitat for species, including the northern spotted owl, that were known to depend on late-successional and old-growth forests.[20] Wind River had been the site of research since 1908 and had been designated specifically for research as an experimental forest since 1932. Comprising less than 1 percent of the total land base of the Gifford Pinchot National Forest, its designation as an experimental forest protected the land for long-term research, not as a source of either additional timber or additional biological diversity for the national forest. By prohibiting some kinds of research, this new designation conflicted with the original purpose of the experimental forest. Although new questions focused in part on how to develop old-growth characteristics in younger stands, the ability to pursue these questions was suddenly limited at Wind River.

This restriction closed the door on many of the research possibilities at Wind River for the silviculturists at the Olympia Lab. Now thinning and other silvicultural treatments in young stands up to 80 years old would be allowed only if the treatments led to the creation of mature or late-successional forest conditions. Although research was still considered "among the primary purposes" of experimental forests in the region, such research projects "would only be considered if there are no equivalent opportunities outside the Late-Successional Reserve." And if approved, the experimental area would have to be surveyed for hundreds of critical species of plants, animals, and fungi, and management plans would have to be designed for their protection.[21]

According to DeBell, the Panther Creek Division would have provided excellent opportunities to study different regeneration systems for multiple purposes. It represented an age class of trees that was typical of much of the federal forests in the Western Cascades, one that forest scientists needed to know more about as they worked to get younger stands into late-successional status. Few lands beyond the federal forests have this age class, yet the Late-Successional Reserve designation eliminated much of the silvicultural work the Olympia Lab might have done at Wind River.

Just as new restrictions limited new work at Wind River, the Olympia Lab's studies at Trout Creek Hill were beginning to bear fruit. Ironically, these studies, begun in the 1970s, examined alternatives to the management regime that was being replaced by the new Northwest Forest Plan. Two studies in particular used the tools of silviculture to examine the benefits of biological diversity in the forest. One examined the silvicultural possibilities of western white pine, one of the fastest growing native conifers in the West. Western white pine had been fairly widespread in the region until the early 1930s when the species was hit by a blister rust, a devastating disease whose alternate host was native gooseberry plants. An experiment established on Trout Creek Hill in the late 1970s tested the possibility of reestablishing this native tree, which was well-suited to a range of sites across the region, including sandy soils, frost pockets, and heavy snow areas. After 16 years researchers successfully established blister rust–resistant western white pine and found that new rust infections appeared to be ebbing.[22] In another Trout Creek Hill study, Constance Harrington, a PNW research forester at the Olympia Lab, examined the relationship of young conifers to understory shrubs, particularly to the nitrogen-fixing shrub ceanothus. She found that ceanothus had a generally positive effect on tree growth, as long as the tops of the growing trees were above the shrubs. Ceanothus, like red alder, had been previously thought of as a weed in Douglas fir plantations and had been removed by cutting and spraying with herbicides.[23]

In a long-term study on the Panther Creek Division begun in 1986, before the Northwest Forest Plan restricted such experiments, DeBell and Franklin worked with district managers to create several experimental harvest patterns in 150-year-old stands of Douglas fir. One block left a number of large, healthy, dominant trees in a so-called shelterwood cut. Twelve years later, Harrington and her colleagues used this block to study how such retention of green trees affected the subsequent natural regeneration of seedlings. They found that the overstory trees had survived—they had not blown down as some had predicted—but only the Douglas fir had regenerated, and these seedlings were growing more slowly than seedlings in nearby plantations. The experiment continues in the 21st century, but these initial results suggested that attaining species diversity within the shelterwood stand would take time.[24]

These studies prompted new questions, but restrictions associated with the late-successional reserve designation eliminated the possibility of testing new treatments at Wind River. Silviculturists would have to find sites in other forests to continue their research. "We are continuing to measure several silvicultural projects we established at Wind River," Harrington explained in 2004, "but due to limitations on the kinds of treatments allowed in Late Successional Reserves, particularly in stands over 80 years old, our future work at Wind River is limited . . . National forests are less interested in tree biology or growth than in the past and less supportive of the silvicultural research projects we have installed."[25]

The Northwest Forest Plan changed forest science in the Douglas fir region in other ways. As harvest levels on federal forests dropped dramatically, there was no longer the need for millions of new seedlings every year. After nearly 90 years of operation, the Wind River Nursery closed in 1997. A ceremony to mark the occasion drew several hundred people from the Forest Service, the timber industry, and the community to pay tribute to the long working relationship between the Forest Service and the local community, a collaboration that produced more than 800 million tree seedlings for reforestation of national forestlands in the Pacific Northwest.[26]

The nursery's closure symbolized a major change in the objectives of Wind River and forest science in the Douglas fir region. But it was not the end of research. Just beyond the boundary of the nursery, in the Thornton T. Munger Research Natural Area, new research was lifting into the canopy, marking the beginning of a new era of science at Wind River.

Chapter Ten
Into the Canopy and Beyond

ᘯ

Once I'm up there, I never want to come down.

—Nalini Nadkarni[1]

The late 20th century saw increasing involvement of the nation's scientists in large-scale environmental studies. Perhaps none was bigger than the study of global climate change and the effect of greenhouse gases on the planet. The new era of science at Wind River would address these global questions at the very top of the forest.

New restrictions within Late-Successional Reserves had made it difficult to continue experiments that involved cutting trees and stalled many of the silviculture experiments that were still planned for Trout Creek Hill. For scientists studying the ecology of old-growth forests, however, it seemed as if the sky was the limit. For most of the 20th century, old forests had been regarded by most forest managers as unproductive graveyards standing in the way of efficient timber production. Now protection of old-growth forests and their associated species was important to the goals of federal forest management. As forest scientists learned more about the structure and function of old forests, they found themselves exploring a new frontier that had long been beyond their reach: the forest canopy.

For the first three-quarters of the 20th century, the leafy top of the forest had been out of reach and mostly out of mind. During that time, the forest canopy was thought of as something like a convertible top on a car: it was either open or closed. When the crowns of those trees grew together, the top was closed, and timber managers knew it was time to thin the stand to get more growth into the remaining wood-bearing trunks.

For generations forest scientists understood that most of the photosynthesis that fed tree growth occurred in the canopy, where sunlight hits the most exposed leaves. As they tested young trees at different spacings, they were in part testing the contribution of different sized crowns to developing wood. But in most studies that considered tree crowns or forest canopies, the treetops were measured from the ground, looking up, or the trees themselves were cut down and dissected into their various components and measured in pieces.[2]

That changed in the 1970s, when a handful of adventurous researchers at the H. J. Andrews Experimental Forest scrambled up into the forest canopy using mountain-climbing gear. Hammering bolts into tree trunks and hoisting themselves up with ropes, they discovered a new ecological frontier, coming eye-to-eye with unknown communities of lichens, insects, birds, and mammals at the very top of the forest.[3] This work inspired canopy studies around the world, and soon researchers, particularly in the tropics, were scaling to new heights.

As research interest grew, so did the inventive ways to get into the canopy. Researchers balanced platforms in the tree limbs, hung swinging rope bridges to go from tree to tree, and experimented with balloons to carry instruments skimming over the world's forest rooftops. The Tarzan-like study of forest canopies grabbed popular attention. A story in a 1991 issue of *National Geographic* featured Nalini Nadkarni, a botany professor at the Evergreen State College in Olympia, exploring the Costa Rican forest canopy with her baby son, Gus, tucked into her backpack.[4]

In the 1980s the Smithsonian Institute installed a construction crane in the Panama rainforest, creating the world's first canopy crane with continuous access to a cylinder of space in the tropical rainforest. Soon afterwards, Jerry Franklin and others began exploring the idea of bringing a canopy crane similar to the one in Panama to the forests of the Pacific Northwest. A canopy crane could be an important tool for science as well as a new source of economic development for a rural community dependent on dwindling supplies of federal timber.

Franklin mentioned the idea to Les AuCoin, then a U.S. congressman from Oregon. He recalled, "AuCoin was really taken with the idea, and he was on the House Appropriations subcommittee that dealt with forest research. The next thing I knew, a subcommittee staff person called me up and asked how much we needed to buy and install a canopy crane."[5]

But finding a site for high-visibility forest research was not easy in the early 1990s. The timber wars had left raw feelings in many timber communities for forest research and anything that appeared to be "tree-hugging" science. The first site proposed for the new crane, on Washington's Olympic Peninsula, was vigorously opposed by the local timber community and its supporters. David Shaw, an Oregon State University (OSU) Extension forester, was working with Franklin at the University of Washington at the time and leading the effort to locate a site for the proposed crane.

"I moved to Forks [Washington] in 1991 and began searching for a location to site the crane," Shaw recalled. "We eventually settled on a location adjacent to the Quinault Research Natural Area, connected to an old-growth forest with a tall, multi-layered canopy. The Environmental Assessment took 16 months to complete, during which time rumors flew in the community that we were looking for more endangered species."

Finally in 1994, the permit application was denied due to threats of violence. The scientists then evaluated Cascade Head, H. J. Andrews, and Wind River experimental forests for possible sites. "We made the final decision to go to Wind River, and with help from the PNW Research Station, the Environmental Assessment took just six weeks to complete," Shaw said.[6]

The old-growth Douglas fir trees at Wind River were not as tall as those on the Olympic Peninsula, but the location had many benefits to recommend it. A potential site was located within the Thornton T. Munger Research Natural Area (RNA), a short distance down an old road where a small creek wound through 500-year-old Douglas firs and western hemlocks. Close to the former nursery fields, the site was level and had access to electricity and nearby office space. And within an hour's drive to Portland International Airport, the site was accessible to researchers flying in from around the world.

The University of Washington, the PNW Research Station, and the Gifford Pinchot National Forest joined forces and established the Wind River Canopy Crane Research Facility in 1995. The crane they purchased, a Liebherr 550 HC, had loomed over urban construction in places ranging from Houston to Hawaii, and most recently had helped construct the San Francisco Public Library. Delivered to Wind River by truck, the 190-ton crane was lugged into the forest and lifted onto a platform of concrete that was 8 feet thick. Once in place, the crane reached up 25 stories through the trees. At the top, a long horizontal arm, called a jib, stretched almost the length of a football field across the top of the forest. The jib swung over a 5.6-acre circle, and from it dangled a steel basket—the gondola. Located in a small, enclosed cab at the top of the crane tower, the crane operator remotely controlled the vertical and horizontal movement of the gondola along the jib. This movement put researchers in touch with every branch and twig within the vast cylinder of space and opened a frontier of filagreed edges previously inaccessible to tree climbers. To reach the cab, the crane operator climbed a series of ladders—a total of 300 rungs—up the tower and into the canopy. For everyone else, the journey was by gondola (fig. 10.1).

Fig. 10.1. The gondola of the Wind River canopy crane swinging over the top of the old-growth forest. (Courtesy of the Wind River Canopy Crane Research Facility Image Archive.)

Researchers, dressed in the requisite climbing harnesses and hard hats, entered the gondola as it rested on the ground. Also in the gondola, the pilot, or arbornaut, radioed information to the crane operator to begin the ascent. The gondola rose slowly above the tangle of vine maple and salal. Rising beside massive tree trunks like an elevator, the gondola was lifted into a landscape of heavy horizontal tree limbs laden with mosses and lichens. Passing through shafts of greenish light, the researchers rose over the stair-step tops of undulating branches and lacy foliage. Then, suddenly, the gondola broke into open sky, atop a rolling, tundra-like field of scruffy snags and lichens.

The gondola swept across the rugged field of treetops and then lowered the researchers into a space between branches to get a closer look at how leaves grow at the very top of trees. The gondola moved in every direction, up and down and across, throughout the three-dimensional cylinder described by the circular swing of the jib. It provided access to the very tips of the branches, through leafy layers, openings, and structural patterns that could not be seen from the ground.*

The canopy was no longer seen just as a roof held up by wood. The Wind River canopy crane opened a world that is just beginning to reveal itself to scientists in the 21st century. As of yet, researchers have not had enough time to fully explore the context and implications of their discoveries. There is no doubt that many of the canopy studies not mentioned in this book will prove to have unexpected importance to future forest research.

One of the first lessons scientists learned from their new vantage point was that the forest canopy was much more complex than they had imagined. It was not just open or closed. The deep narrow crowns and undulations in the outer canopy created a complex surface with eight times more leaf area than the ground below (fig. 10.2).[8] Gaps in that outer canopy opened to layers of lower canopies, creating

---

* At 250 feet tall with a reach of 279 feet, the Wind River crane is one of the largest canopy cranes now in operation in the world. Since its installation, several others have been erected in cities around the world, including Australia, Venezuela, Switzerland, Japan, and Borneo, as well as two in Germany and two in Panama.[7]

Fig. 10.2. Looking down at the treetops from the Wind River canopy crane. (Courtesy of the Wind River Canopy Crane Research Facility Image Archive.)

stacks of microclimates and microhabitats. Researchers referred to the layered structure as architecture and found plants and animals using different parts of that architecture for different purposes. Throughout the cylinder of space defined by the crane, scientists took stock of what they found, from dwarf mistletoe to lichens and from spiders to birds.

A notable characteristic of Douglas fir that caught the attention of researchers throughout the century was that these trees grow to be very big and very old. What were the secrets to living old and well in Pacific Northwest forests? For one, old-growth Douglas fir trees were tall; they dominated the upper canopy. And they persist; their dominance in the upper canopy is likely to last for centuries. In addition, old-growth Douglas firs have deep crowns and an open branching pattern that allows sunlight to filter through the canopy and bathe foliage in light. This is not true of younger Douglas fir that seem to race to the sky and then close the canopy to incoming light.

Like some people, as the trees grow older they develop more in the middle. Trees develop architectural characteristics that help resist disturbance. Among those characteristics are epicormic branches, secondary limbs that erupt from dormant buds buried in the trunk or other branches. Exploring the canopy, scientists found that as much as 70 percent of all the foliage on old-growth Douglas fir trees may be found on epicormic branches.[9]

A tree needs water and light to make food in a process known as photosynthesis. During times of drought, the longer a tree can pump water up into its leaves, the longer it can make food and grow. Scientists using the Wind River canopy crane examined conifer needles in different layers in the canopy and found more active photosynthesis in the brightly lit treetops than at the bottom or in younger saplings that are often in the shade. In some cases, they found photosynthetic rates at the top of the towering Douglas fir trees to be higher than rates measured for young Douglas fir trees growing in plantations. Such high rates of photosynthesis were sustained even at the end of the yearly summer drought, during which it had been thought that photosynthetic activity would be greatly reduced due to water stress. Research at the crane showed conclusively that photosynthesis occurs year- round in these conifers. Although photosynthesis slows down in the winter to 40 percent of the rate during the summer, scientists found that it was not winter cold but a lack of sunlight on dark winter days that slowed the process.[10]

As a tree ages, its wood production slows. The earliest research at Wind River by E. T. Allen and T. T. Munger confirmed this, which led to interest in managing younger trees for maximum wood production. Researchers assumed that big old trees allocated all their photosynthetic effort to respiration, just to keep their cells alive but not to add significant new growth. But in the late 1980s, Forest Service research forester Michael Ryan measured respiration in lots of forest types of different ages and found that although old trees do respire more than young ones, the difference was not statistically significant. Barbara Bond, a professor of forest physiology at OSU, joined Ryan to pursue the question at the canopy crane and in young stands within the experimental forest. They found that big old trees photosynthesized less than young trees, perhaps because of the difficulty of getting water to the top of tall trees. But the difference they found in water use and photosynthesis was not big enough to explain the growth reduction. Where was the carbon going that was fixed by photosynthesis but not showing up in new growth? Where was the missing sink? Was the extra carbon going underground?[11]

The breathing forest, particularly its role in the uptake and release of carbon, had become a topic of increasing interest following the Kyoto Convention in 1994, when most nations agreed to reduce emissions of greenhouse gases responsible for global warming. Carbon dioxide is foremost among those gases that increase the atmosphere's ability to trap heat. Carbon is absorbed and stored in vegetation as living plants or the coal and oil formed from ancient plants, and is released into the atmosphere as carbon dioxide when vegetation is cleared or burned. But carbon storage and release is not a simple matter of moving molecules from one place to another. Each year about 8 billion metric tons of carbon are dumped into the atmosphere from burning fossil fuels and deforestation. About half that amount gets absorbed by forests, grasslands, and the waters of the oceans, slowing carbon buildup in the atmosphere and delaying changes to climate.[12] Carbon dioxide's contribution to the greenhouse effect created puzzling questions for science. What were the sources where carbon was being pumped into the atmosphere? And what were the sinks where carbon was absorbed from the atmosphere?

Incorporated as one of several sites in the National Institute of Global Environmental Change, the crane at Wind River was instrumented to measure fluxes of carbon dioxide and water vapor using a method called eddy correlation. This method, also known as eddy flux, measures the fluxes of gases that occur above growing plants, such as in a forest canopy. As air moves across the canopy, invisible swirling eddies carry gases such as carbon dioxide and water vapor upward and downward. The amount of gas carried in these eddies of air during a given time period is called a flux. To measure fluxes of carbon dioxide and water vapor in the forest, scientists installed highly precise instruments along the crane tower that continuously measured wind speed and the amount of water vapor and carbon dioxide in the air (fig. 10.3). Depending on the concentration of water vapor and carbon dioxide and the speed of the air moving up and down in eddies, scientists could measure the net amount of carbon dioxide absorbed by the forest.

Six years of continuous eddy flux measurements showed that old-growth forests could absorb a small amount of carbon as a sink. However, significant variation from day to night, season to season, and

Fig. 10.3. Eddy correlation instruments attached to the tower of the Wind River canopy crane. (Courtesy of the Wind River Canopy Crane Research Facility Image Archive.)

year to year suggested that at times the forest absorbed a significant amount of carbon and at other times it leaked slight amounts of carbon into the atmosphere.[13] In a related study, Mark Harmon, a professor of forest science at OSU, examined carbon storage in trunks, soil, roots, and coarse woody debris, using long-term data recorded from permanent sample plots at Wind River. Harmon found that over the course of 50 years the Wind River forest was a slight sink. Concealed in Harmon's long-term study was the small yearly variation recorded in the eddy correlation data that showed how carbon responds to short-term climate variations. Harmon's long-term study smoothed out short-term variations to estimate the role of the forest in the global carbon exchange over time.[14]

Despite its soggy reputation, the Pacific Northwest can be virtually rainless for several months during the summer. A tree must be able to take up water in order to use carbon for growth and respiration. During long, dry summers, conifers of the Pacific Northwest forests tend to bring water up from deep roots that are as much as six feet underground to fine roots near the top of the soil. This hydraulic

Fig. 10.4. Rick Meinzer, PNW research ecologist, in the gondola cutting foliage to determine water status, 2002. (Courtesy of the Wind River Canopy Crane Research Facility Image Archive.)

lift fascinated Wind River scientists, including Frederick Meinzer, a PNW research ecologist working at the crane facility. Meinzer and his colleagues found that Douglas fir roots were able to redistribute water from deeper, wetter parts to shallower, drier parts of the soil. Without this redistribution of water to the roots in shallow soil layers, failure of the root-water transport system could occur, cutting off water and nutrient uptake to the tree (fig. 10.4).[15]

Drought is not the only problem Northwest forests face. Parasites can stress forests, and one organism in particular, hemlock dwarf mistletoe, is widespread in the western hemlock of the Douglas fir region. With the crane, scientists were able to examine this parasitic plant where it grew, high in the branches of western hemlock. Each fall, hemlock dwarf mistletoe shoots out its seeds, sometimes with enough force to travel 50 feet. Where the seeds take hold, mistletoe grows and can stunt the growth of the host tree and deform its branches into bristling growths called witches' brooms. In some cases the tree is killed.

While working as a research scientist at the crane facility, Shaw examined the distribution and abundance of hemlock dwarf mistletoe using the permanent plots in the research natural area. He found that the seed-producing aerial shoots were most common high in the canopy in areas with plenty of sunlight where more spaces between trees and crowns made it easier for hemlock dwarf mistletoe to spread through the old-growth forest. Because hemlock dwarf mistletoe attacks only hemlocks, the presence of unsusceptible Douglas firs and cedars blocked some of the shooting seeds from landing on their intended targets.[16] He found the parasitic plant in only a few locations in the younger forest stands at the Panther Creek Division, along a creek that had escaped the fires that burned most of this part of the forest in the mid-1800s. Shaw speculated that the replacement stand of almost pure Douglas fir stopped the spread of hemlock dwarf mistletoe, and the few unburned riparian areas maintained a refuge for the parasite.[17]

The maintenance of biological diversity had been a key goal written into the Northwest Forest Plan, and the canopy crane provided new opportunities to study the diversity of the forest ecosystem. Researchers found that lichen diversity increased but moss diversity decreased with increasing canopy height. And in the upper six feet of the canopy researchers found a lichen community associated only with dead tops and roosting posts of birds.[18] Birds too were found to stratify themselves within the canopy in arrangements that can shift with the seasons.[19] Researchers found that arthropod diversity was determined largely by the species of the host tree; Douglas fir had the greatest number of distinct kinds of spiders and mites.[20] These studies linked biological diversity to the complex canopy structure of old forests and trees of different ages and species. By eliminating this complexity, forests managed for single-aged, short-rotation Douglas fir would limit the possibility for such biological diversity.

The end of the 20th century saw a new appreciation for some of the oldest tools of forest science: permanent sample plots. Since 1910, when Munger established the first permanent sample plots at Wind River and beyond, generations of forest scientists have devoted attention and diligence to periodic measurements, data analysis, and records maintenance. Considering the representative nature of permanent plots across the nation, and the collective effort and results they portray, Boris Zeide, professor of forestry at the University of

Arkansas, suggested that "permanent plots are more than a method of science; they may be viewed as a cultural phenomenon. Permanent plots deserve to be treasured as living national monuments."[21]

The original permanent sample plots in the Pacific Northwest were designed to measure timber growth and yield over time. The oldest of these plots established by Munger in 1910 were remeasured for the fourteenth time in 2002.[22] Throughout the 20th century, as new questions arose, long-term data have offered new insights. For example, recent remeasurements of the old stands within Thornton T. Munger RNA were compared with similar data from the Fraser Experimental Forest in Colorado to learn how various old-growth forest ecosystems store or release carbon. And two new sets of permanent plots were established within riparian areas at Wind River, places that are integral to the function of healthy forest and stream ecosystems. These streamside plots will provide long-term data on cover and shade to the stream and coarse woody debris that form habitat for many fish and wildlife species.

Gap studies contribute to another long-term data set at Wind River that has helped winnow the relation between small disturbances and long-term dynamics within the forest. Gaps are a feature of mature forests, where trees have fallen and opened a skylight in the forest canopy. Ecologists identified these gaps as potential hot spots for species diversity, nutrient cycling, and forest productivity. Silviculturists questioned if gaps could speed up the development of late-successional characteristics in younger forests. In a study at Wind River and H. J. Andrews experimental forests begun in 1990 (before the Northwest Forest Plan), Thomas Spies, research ecologist at the PNW Research Station lab in Corvallis, pursued these and other questions. Spies, with help from the local ranger districts, created different sized gaps, openings in the forest up to 150 feet across, to examine ecosystem responses to small disturbances in forests. In doing so, he created a site for long-term research, with a thorough record of existing conditions to use as a baseline for future studies. Spies found that Douglas fir and other conifers regenerated in these columns of light, but survival depended on many factors, including differences in light, moisture, temperature, seed availability, and forest floor debris. He found that gaps aboveground indicated gaps belowground, where root density was reduced and soil moisture increased.[23] In a way, gaps were analogous to patch cutting, one of many alternatives to clearcutting explored by Burt Kirkland and Axel Brandstrom more than 60 years earlier. Although designed as ecosystem studies, these modern gaps are also useful to silviculture, as they provide information about how understory plants react to logging disturbance and how gap sizes can affect regeneration.

As scientists learned more about the structure and function of mature forests, they began to question how that structure and function would change as the forest aged. One way to examine this was to use a chronosequence, a series of stands of different ages thought to represent how a forest develops through time. Conveniently, Wind River and its environs proved to be a good place for such studies, with forests in a patchwork of different ages created by fires, harvests, and experiments. Forest stands ranging from a very young age to nearly 1,000 years old provide a laboratory that substitutes different aged stands for different amounts of time.

Research at Wind River now spans from branch tips to global climate. In the 21st century the experimental forest and canopy crane facility will play a role in the National Ecological Observatory Network (NEON), a continental-scale research network planned by the National Science Foundation. NEON will connect social scientists and educators with ecologists and physical scientists to answer large-scale environmental questions with state-of-the-art computational, analytical, and modeling capabilities.

Increasingly, forest experiments span large scales that extend beyond the boundaries of the experimental forests. Two studies in particular that examine silviculture and ecosystem science represent new directions forest science has taken since the end of the 20th century. Although neither study is conducted at Wind River, each involves a core of scientists who have worked at Wind River and builds on knowledge developed at the experimental forest.

A large-scale replicated experiment on the effects of timber harvesting in west-side Douglas fir forests is underway in the Wind River area and elsewhere around the region. Called DEMO (Demonstration of Management Options), this long-term experiment is an ecosystem-based study of alternatives to clearcutting. The primary emphasis is the effects on biological diversity of retaining and enhancing old-growth characteristics after harvest. The experiment compares six different patterns of retaining trees after harvesting timber and how each treatment accelerates the recovery of plants, animals, and ecological features of mature and old-growth forests. The study, replicated in large blocks across the Douglas fir region, examines patterns of retaining standing trees reminiscent of staggered settings of harvest sites that Leo Isaac described in 1943.[24] The blocks retained in the staggered setting approach were meant to provide fire protection and a source of seed, while the DEMO retentions are meant to enhance biological diversity and rebuild old-growth structure. Just as in the modern gap study, DEMO draws from concepts that are more than 60 years old, concepts that are revisited in a new context of public expectations for the federal forests. Timber production was not a primary objective of DEMO, but the study included a public opinion survey that found a consensus in favor of retaining some trees on the landscape rather than prescribing clearcutting and plantation silviculture.[25]

Beyond any scientific debate about clearcutting timber lay the fact that many people object to the look of a stump field. Aesthetics has become increasingly important to forest management practices as a non-timber value that is highly visible to the public and one that could cause trouble for forest managers if it is ignored. A second study in the region addressed aesthetics and the goals of economic timber production on a site in the state-owned Capitol Forest, near Olympia. Capital Forest was a silvicultural examination of alternatives to clearcutting. Here scientists addressed timber production through a second and possible third harvest, comparing clearcuts with various selection cuts and extended rotations, and documenting public reactions to the aesthetics of various harvest treatments.[26]

In the 21st century private timberland owners have retooled their mills to handle smaller trees and are managing their forests on harvest rotations of 40 years or less. But growing a forest is different than growing timber. Federal forest managers are managing forests with longer rotations and thinnings that leave behind elements of a remnant forest and encourage old-growth characteristics to develop. One of the most difficult characteristics to build into younger forests is the accumulation of large biomass, both living and dead.

According to Andrew Carey, decadence—dead wood and the process of decomposition—is one of the most fundamental forest-building processes and the one that is most negatively affected by years of single-use management for timber production.[27] Decadence, dead wood, and what Munger had described as the menace of murderous snags were systematically removed in the 20th century. Now they are being systematically returned in the 21st century.

Such changes in scientific assumptions should not be dismissed as stupidity in the past, nor as folly in the future. Science, if practiced honestly, is an adaptive process and reacts to social issues. Knowledge changes with experience, and the kind of knowledge pursued by science is affected by society's values. It would be far worse for science to become satisfied with its apparent success, stop testing its experience, and stop measuring change.

Consider the Wind River Arboretum. In the late 1980s Roy Silen remeasured the collection of imported trees Munger had planted in 1912 and found lessons of both success and failure. Most of the broad-leafed trees had failed early. Hardier introduced species were damaged over time by disease, insects, or the sudden ravages of a single storm. Even the European larches that showed promising growth and vigor for more than a half century were beginning to decline by century's end. But according to Silen, no finding was more important than the success of Pacific Northwest native firs, pines, hemlocks, and cedars that far outgrew their genetic relatives from other regions. In addition, when introduced to other lands, some of these trees, particularly Douglas fir and Sitka spruce, outgrow the native species in these other lands. The Pacific Northwest, as it turns out, grows outstanding trees, particularly its native trees.[28]

The Wind River Arboretum remains a popular site for tourists and visiting scientists. Currently researchers are using it to tease out the factors that have restricted nonnative conifers and to build knowledge about biological invasions. "The long time span required to express a species mismatch to the environment becomes clarified by this seven-decade record," Silen said, reflecting on the value of data recorded at the arboretum.[29]

In the century since the first forest scientists arrived at Wind River, forest research has changed in scale and focus from regeneration and timber production to biological diversity and forest dynamics. Many would argue that little has really changed, that the questions pursued by Munger, Hofmann, and Isaac were essentially the same as those pursued by Silen, DeBell, and Franklin. Others might suggest that the differences come down to changing ideas about economics and the environment, more concern for aesthetics, more pressure from politicians, and new concerns about a changing climate. At times throughout the century, it may have seemed that science could not see the forest for the trees, or the trees for the forest. Conflicts sometimes kept silviculture and ecosystem studies on separate paths. By the beginning of the 21st century, those paths may be converging. From years of research on how to grow trees, it is now possible to ask how to grow a forest.

Afterword

Looking into the 21st Century

ℒ

*The lighter it grows around us, the more unknown things
become apparent, and it is a sure sign of shallowness if
anybody believes he knows it all.*

— Heinrich von Cotta[1]

In 1902, the year of the Yacolt fire, Bernhard Fernow translated the
preface of an historic text on forestry that had been written by Fernow's
countryman, Heinrich von Cotta, in 1816. "Each generation of man
has seen a smaller generation of wood," Cotta wrote. "We have now a
forestry science because we have a dearth of wood."

Fernow recognized Cotta's preface as a cautionary note for forest
scientists at the beginning of the 20th century, and it serves the same
purpose now at the beginning of the 21st. Cotta was among the first
to scientifically investigate changing forests, examine the decline of
the European forest, and consider ways to regenerate it. Cotta saw
shortcomings in the science of forestry, even in its earlier iterations.
He warned that the forest was far more complicated than imagined
and urged forest scientists to know the forest as something far more
vast and complex than the sum of its parts. Cotta recognized that
scientific understanding of the forest will change at various points in
time, at various locations, and through the eyes of various people. He
wrote, "Forestry is based on the knowledge of nature, the deeper we
penetrate its secrets, the deeper the depths before us."[2]

A century after Cotta published his treatise, T. T. Munger and
others were composing a new chapter in forest science in the Wind
River forest. It is tempting to think that when Munger was planning
the arboretum at Wind River in 1911, he might have reviewed Cotta's

1811 establishment of one of the world's first scientific arboreta in Tharandt, Germany.[3] Munger might have read about the research conducted in the 19th century in experimental forests in Germany, Prussia, and France, forests that would become models for Wind River and the 20th-century system of American experimental forests.

Munger and his Wind River colleagues faced a different kind of forest than Cotta had described in Germany in the early 1800s. The Douglas fir forests of the Pacific Northwest revealed ancient memories of wind and fire in patchwork patterns across the landscape. Science had to learn to truncate time. At the beginning of the 20th century, with concern about the cut-and-run fate of the nation's last great stands of timber, science developed an agricultural approach that sought to plant trees as crops and eliminate fire and pests from the forest. At midcentury, a call for raw materials to support a national building boom prompted science to develop a technological approach to maximize young timber production in place of older forests. At the end of the century, public concern over the loss of old-growth forests turned the technological approach on its head, and science pursued ways to recreate old-growth forests. The silvicultural tools that sped trees from seed to harvest were now being used to speed the creation of old-growth characteristics in younger forests. Each approach made sense in its own historical context. But clearly throughout the century, no single approach to forest research and management in any one time was able to provide for the needs of a changing society and a changing forest over time. Forests and forest research constantly outgrow one-size-fits-all scientific approaches.

The challenge of every generation is to see the larger context of time. Assumptions, findings, and society's expectations all change in much less time than it takes to grow a forest. Will the management of federal forests for old-growth characteristics seem like a single-minded prescription to forest scientists a hundred years from now? Where will timber come from in the 22nd century? What will be the questions, assumptions, and essential tools of science in the Pacific Northwest?

Forest science at Wind River has been a century-long conversation among different people with different ways of seeing the forest. Each contributed important perspectives to forest science in the Douglas fir region, each influenced by the context of the times. Munger, an easterner trained in European traditions, challenged the cut-and-run approach of the timber industry at the beginning of the 20th century. Leo Isaac devoted his career to understanding how Douglas fir grows. Dean DeBell and Jerry Franklin examined the old-growth forest from different angles, using complementary lenses, and opened the way for more integration of sciences. Canopy studies further broadened the emphasis from board-feet calculations to analyses of forest structure and function from roots to treetops. Carbon studies extended the scale of research from molecules to global climate patterns. For 100 years, scientists have come to Wind River and have seen the forest in different ways. Their work, as much as the trees, defines the experimental forest.

Throughout the generations of research at Wind River, scientists encountered unexpected events that shook their assumptions and created scientific confusion. Long-term research and a forest dedicated to collaborative, long-term experiments provide perspective when these surprises occur. Many of the permanent plots Munger established in 1910 are still measured in the 21st century and continue to add new knowledge to how forests grow through time. The arboretum that Munger created is now overshadowed by the towering native Douglas firs that have grown up at its edge.

Wind River is a place where people have come to study the forest for more than a hundred years. It is not a school of thought. There is no overriding worldview; there is not even a recognizable unit that anyone could call "the Wind River group." The story of Wind River is all the more resonant because there is no defining culture that limited the experience of people who worked there. The Wind River story moves through time by the force of discovery, disagreement, debate, and changing assumptions. It is the way science progresses. It is all the more important these days that readers—scientists, students, and the public—understand that science is not a perfect marriage of social values and unvarnished truth. We witness research at odds with itself every day. Disagreements and debates test the strength of ideas but rarely prove the truth of anything. We hope that our readers gain a clearer understanding of how scientific knowledge changes through time, by the back-and-forth pressure of scientific debate and by the slowly evolving evidence that only long-term studies can provide.

The Wind River Experimental Forest is a place, a process, an experiment, a lesson. Throughout the 20th century, generations of researchers developed new ideas about the forests in the Pacific Northwest. Assumptions and values changed, sometimes reversing themselves in a very short time. One hundred years of science at Wind River shows what each generation of scientists learned and accomplished with the passage of time. Through the ebb and flow of funding and changing research questions, the value of the place increases with time. Forest science has gone through many iterations in the last hundred years. There have always been dissenting voices. Debates continue. That is the strength of science and the underlying lesson from the Wind River story.

The next century of forest science will face new challenges of balancing timber production, forest sustainability, and social values. Solutions to these challenges will require multiple approaches drawn from both silviculture and ecosystem studies. "Things look very differently in the forest from what they do in books," Cotta wrote two centuries ago.[4] It is more important to envision the future forest on the ground than in a conference room. Therefore, Wind River will continue to be important as a place of discovery where real trees grow through time.

Wind River Experimental Forest Timeline

1816   Heinrich von Cotta publishes *Advice on Silviculture*.

1886   Bernhard Fernow becomes division chief, Bureau of Forestry, Department of Agriculture.

1891   Congress gives president authority to establish forest reserves in the public domain.

1897   President Cleveland establishes Mount Rainier Forest Reserve.
Congress passes Organic Act.

1898   Gifford Pinchot becomes division chief, Bureau of Forestry, and establishes Section of Special Investigations to gather scientific data on forest reserves.

1901   Raphael Zon joins Gifford Pinchot at Bureau of Forestry.

1902   Horace Wetherell begins work at Wind River, Mount Rainier Forest Reserve.
Yacolt fire burns in southwest Washington.

1903   E. T. Allen writes "Red Fir in the Northwest," the first definitive work on Douglas fir in the Pacific Northwest.

1905   President Theodore Roosevelt moves forest reserves from the Department of Interior to the Department of Agriculture, thereby establishing the Forest Service.

1906   First federal timber sale purchased in the Wind River valley.

1907   Mount Rainier Forest Reserve renamed the Columbia National Forest.

1908   T. T. Munger heads west to study encroachment of lodgepole pine in ponderosa pine forests.
Forest Service establishes first research experiment station at Fort Valley, Arizona.

1909   Wind River Nursery established.

1910   Munger establishes first permanent plots in Douglas fir stands.

1912   Wind River Arboretum established.
Wind River Experiment Station established.

1913   Julius Hofmann becomes director of Wind River Experiment Station.

1914   World War I begins.
1915   Munger establishes first Douglas fir plantations for heredity study.
       Forest Service establishes independent Branch of Research.
1918   World War I ends.
       Hofmann establishes Wind River transect.
1919   Hofmann establishes precommercial thinning plots at Martha Creek.
1923   Hofmann establishes importance of relative humidity measurements to fire predictions.
1924   Munger heads newly established Pacific Northwest Forest and Range Experiment Station in Portland, Oregon.
       Leo Isaac arrives at Wind River, replacing Hofmann as director of the Wind River Experiment Station.
       Congress passes Clarke-McNary Act.
1925   Isaac establishes spacing study.
1926   Chief of Forest Service establishes Wind River Natural Area.
       Munger establishes Regional Races of Ponderosa Pine study.
       Isaac, using a kite, tests how far Douglas fir seeds can fly.
1928   Congress passes McSweeney-McNary Act.
1929   Stock market crashes.
1930   Walter Meyer and Richard McArdle publish "The Yield of Douglas Fir in the Pacific Northwest," more commonly known as Bulletin 201.
1932   Wind River Experimental Forest established.
1933   Civilian Conservation Corps (CCC) builds Camp Hemlock at Wind River.
1935   Isaac publishes "The Life of Douglas Fir Seed in the Forest Floor," debunking theory of buried seed in the duff.
       William Morris begins fire moisture studies.
       CCC builds Trout Creek dam.
1936   Burt Kirkland and Axel Brandstrom publish *Selective Timber Management in the Douglas Fir Region*.
1938   First commercial thinning study of mature Douglas fir established in Panther Creek Division.
1942   United States enters World War II.
1943   Isaac publishes *Reproductive Habits of Douglas Fir*.

1945   World War II ends.

1946   Roy Silen and Robert Tarrant join Forest Service.
Gifford Pinchot dies.

1947   Permanent plots established in Wind River Research Natural Area (RNA).

1949   Columbia National Forest renamed Gifford Pinchot National Forest.

1950   Experimental thinning study to control root rot in Panther Creek Division begins.

1953   George Staebler establishes thinning study at Planting Creek.

1956   Isaac publishes *The Place of Partial Cutting in Old Growth Stands of the Douglas Fir Region*, refuting the call for selective timber management.
Tarrant begins Alder Conifer study in Douglas fir–alder plantation.

1960   Congress passes Multiple Use–Sustained Yield Act.

1962   Rachel Carson publishes *Silent Spring*.
Americans land on the moon.

1964   Nitrogen fertilizer studies begin.

1969   Congress passes National Environmental Policy Act.

1972   Jerry Franklin and Dean DeBell present paper titled "Effects of Various Harvesting Methods on Forest Regeneration."

1973   Congress passes Endangered Species Act.

1975   Munger dies. Wind River RNA renamed Thornton T. Munger RNA. Study plans for Trout Creek Hill finalized and first timber sale made.

1976   Congress passes National Forest Management Act.

1980   Work begins on new Wind River Experimental Forest management plan.

1981   Franklin et al. publish "Ecological Characteristics of Old-growth Douglas-fir Forests."

1983   Studies begin on wildlife and vegetation of unmanaged Douglas fir forests.

1986   DeBell and Franklin establish shelterwood study in Panther Creek Division.

1987   Wind River Experimental Forest management plan completed and signed.

1990   Thomas Spies establishes Gap study.

1993   Northern spotted owl listed as threatened.

1994   Northwest Forest Plan finalized.

1995   Pacific Northwest Research Station, Gifford Pinchot National Forest, and University of Washington establish Wind River Canopy Crane Research Facility.

1997   Wind River Nursery closed.

2005   10th annual science conference held at Wind River Canopy Crane Research Facility.

2007   WREF chosen as core domain site as part of NEON.

Notes/References

Introduction

1. Margaret E. Felt, "Yacolt! The forest that would not die" (Washington Department of Natural Resources, Olympia, WA, 1977), 23.
2. Ibid.
3. Rhonda Bartosh, "1902 DNR Yacolt Burn history, passport to historical southwest Washington, project summary," in *Weather and fire* (Washington Department of Natural Resources, Olympia, WA, 1995).
4. Stephen J. Pyne, *Fire in America: A cultural history of wildland and rural fire* (Princeton, NJ: Princeton University Press, 1982), 2.
5. Donald Culross Peattie, *A natural history of western trees* (Boston: Houghton Mifflin, 1953), 169–81.
6. Meriwether Lewis, William Clark, and members of the Corps of Discovery, *The Lewis and Clark journals: An American epic of discovery; The abridgement of the definitive Nebraska edition*, ed. Gary E. Moulton (Lincoln, NE: University of Nebraska Press, 2003), 328.
7. John Muir, "Douglas squirrel," in *The mountains of California* (New York: Century, 1894), 252–71.
8. See note 5 above.
9. Carlos A. Schwantes, *The Pacific Northwest: An interpretive history* (Lincoln, NE: University of Nebraska Press, 1996), 181.
10. Robert E. Ficken, *The forested land: A history of lumbering in western Washington* (Durham, NC: Forest History Society; Seattle and London: University of Washington Press, 1987), 13.
11. Jamie Tolfree, "History of the Wind River Lumber Company in the Wind River valley, part one," *Skamania County Heritage*, bulletin 13, no. 3 (March 1985), 4.
12. Ficken, *Forested land*, 4–5.
13. William G. Robbins, *Landscapes of promise: The Oregon story, 1800–1940.* (Seattle: University of Washington Press, 1997), 235.
14. Henry Gannett, *The forests of Washington: A revision of estimates*, USDI USGS Professional Paper No. 5, Series H, Forestry, 2 (Washington, DC: Government Printing Office, 1902), 7.
15. Jamie Tolfree, "History of the Wind River Lumber Company in the Wind River valley, part two," *Skamania County Heritage* bulletin 13, no. 4 (March 1985), 7.
16. Felt, "Yacolt!" 16.
17. Paul W. Hirt, *A conspiracy of optimism: Management of the national forests since World War Two* (Lincoln, NE: University of Nebraska Press, 1994), 29–31.

18. William G. Robbins, *Lumberjacks and legislators: Political economy of the U.S. lumber industry, 1890–1941* (College Station, TX: A&M University Press, 1982), 26.

19. Charles S. Cowan, *The enemy is fire* (Seattle: Superior, 1961), 37.

20. Gifford Pinchot, *Breaking new ground* (New York: Harcourt and Brace, 1947), 281.

Chapter One

1. Bernhard E. Fernow, *A brief history of forestry: In Europe, the United States and other countries* (Toronto, ON: University Press, 1911), 1.

2. Bernhard Fernow, *Report upon the forestry investigations of the USDA, 1877–1898* (Washington, DC: Government Printing Office, 1899), 6.

3. Heinrich von Cotta, "Cotta's preface," *Forest History Today* (Fall 2000): 27.

4. Marion Clawson and Roger Sedjo, "History of sustained yield concept and its application to developing countries," in *History of sustained-yield forestry: A symposium; Western Forestry Center, Portland, Oregon, October 18–19, 1983* (Santa Cruz, CA: Society, c1984), 3–9.

5. Andrew Denny Rodgers, *Bernhard Eduard Fernow: A story of North American forestry* (Princeton, NJ: Princeton University Press, 1951), 74–75.

6. Fernow, *Report upon the forestry investigations*, 17.

7. Fernow, *Brief history of forestry*, 484.

8. Ibid., 458.

9. Ibid., 487.

10. Ibid., 487.

11. William G. Robbins, *Landscapes of conflict: The Oregon story, 1940–2000.* (Seattle: University of Washington Press, 2004), 289.

12. Richard Allan Rajala, *Clearcutting the Pacific rain forest: Production, science, and regulation* (Vancouver, BC: University of British Columbia Press, 1998), 88.

13. William G. Robbins, *American forestry: A history of national, state, and private cooperation* (Lincoln, NE: University of Nebraska Press, 1985), 17.

14. Char Miller, *Gifford Pinchot and the making of modern environmentalism* (Washington, DC: Island Press/Shearwater Books, 2001), 155.

15. Pinchot, *Breaking new ground*, 32 (see introduction, n. 20).

16. Shirley W. Allen, "We present . . . E. T. Allen," *Journal of Forestry* 48, no. 11 (1950): 754.

17. E. T. Allen, "Red fir in the northwest" (report on file, USDA Forest Service, Pacific Northwest Research Station, Portland, OR, 1903), 1, 86.

18. See note 16 above.

19. Rodgers, *Bernhard Eduard Fernow*, 379.

20. Obituary for E. T. Allen, *Journal of Forestry* 40, no. 7 (1950): 574–75.

21. Schwantes, *Pacific Northwest*, 219 (see introduction, n. 9).

Chapter Two

1. Thornton T. Munger, "Forest research in the Pacific Northwest: An interview conducted by Amelia R. Fry" (Forest History Society and Hill Family Foundation, Berkeley, CA, 1967), 42–43.

2. Pinchot, *Breaking new ground*, 152 (see introduction, n. 20).

3. Robbins, *Lumberjacks and legislators*, 21 (see introduction, n. 18).

4. Nancy Langston, *Forest dreams, forest nightmares: The paradox of old-growth in the inland west* (Seattle: University of Washington Press, 1995), 136.

5. T. J. Starker to Robert F. Tarrant, 30 September 1975, on file, USDA Forest Service, Pacific Northwest Research Station, Portland, OR.

6. Munger, "Forest research in the Pacific Northwest," 36.

7. Ibid., 42.

8. George B. Sudworth, *Forest trees of the Pacific slope* (Washington, DC: Government Printing Office, 1908), 15.

9. Munger, "Forest research in the Pacific Northwest," 48.

10. Ibid., 49.

11. Review of *On the threshold*, by Theodore T. Munger, *Century* 23, no. 1 (November 1881), 154.

12. Munger, "Forest research in the Pacific Northwest," 45.

13. Ibid., 44.

14. E. T. Allen, *Practical forestry in the Pacific Northwest* (Portland, OR: Western Forestry and Conservation Association, 1911), 3.

15. Thornton T. Munger, *Growth and management of Douglas-fir in the Pacific Northwest*, USDA Forest Service Circular 175 (Washington, DC: Government Printing Office, 1911).

16. Munger, "Forest research in the Pacific Northwest," 44–45.

17. Munger, *Growth and management*, 15, 23.

18. Munger, "Forest research in the Pacific Northwest," 45.

19. See note 15 above.

20. J. K. Agee and F. Krusemark, "Forest fire regime of the Bull Run watershed, Oregon," *Northwest Science* 75, no. 3 (2001): 292–306.

21. "Columbia-Nurseries-Wind River Annual Report 1912" (on file, Mt. Adams Ranger Station, Gifford Pinchot National Forest, Trout Lake, WA).

22. Cheryl Mack, "Archaeological test excavations and significance evaluation of four sites in the Wind River Nursery Fields, Gifford Pinchot National Forest" (report on file, Heritage Program, Gifford Pinchot National Forest, Vancouver, WA, 1999), 19.

23. See note 5 above.

24. Rick McClure and Cheryl Mack, *For the greatest good: Early history of the Gifford Pinchot National Forest* (Seattle: Northwest Interpretive Association, 1999), 24.

25. See note 5 above.

26. Ibid.

27. McClure and Mack, *For the greatest good*, 10.

Chapter Three

1. Pinchot, *Breaking new ground*, 308 (see introduction, n. 20).

2. Ibid., 308.

3. Ibid.

4. Norman J. Schmaltz, "Forest researcher, Raphael Zon," *Journal of Forest History* (January 1980), 26.

5. Robbins, *American forestry* (see chap. 1, n. 13).

6. Rodgers, *Bernhard Eduard Fernow*, 409 (see chap. 1, n. 5).

7. Schmaltz, "Forest researcher, Raphael Zon," 27.

8. Susan Deaver Olberding, "Fort Valley: The beginning of forest research," *Forest History Today* (Spring 2000): 9.

9. Munger, "Forest research in the Pacific Northwest," 63 (see chap. 2, n. 1).

10. June H. Wertz, "A record concerning the Wind River Forest Experiment Station July 1, 1913–June 30, 1924 and the Pacific Northwest Forest and Range Experiment Station July 1, 1924–December 30, 1938 with supplements through 1943" (report on file, USDA Forest Service, Pacific Northwest Research Station, Portland, OR, 1940), 2.

11. Julius Hofmann, "Humidity of atmosphere and behavior of fire" (speech, Pacific Logging Congress, Portland, OR, 1923).

12. Charles S. Cowan, "Fire protection in the Pacific Northwest: An oral history interview," interview by Elwood Maunder, 30 October and 2 November 1957, rough transcript, Forest History Society, Durham, NC.

13. A. L. Westerling, H. G. Hidalgo, D. R. Cayan, and T. W. Swetnam, "Warming and earlier spring increase western U.S. forest wildfire activity," *Science* 313, no. 5789 (2006): 940–43.

14. Leo A. Isaac and E. G. Dunford, "Twenty years of natural regeneration on a Douglas fir cut-over area: Fourth progress report—Camp 8 plots" (report on file, USDA Forest Service, Pacific Northwest Research Station, Forestry Sciences Laboratory, Corvallis, OR, 1936).

15. Ibid.

16. Dean DeBell, "Historic transect across the Wind River Valley" (report on file, USDA Forest Service, Pacific Northwest Research Station, Corvallis, OR, 2003).

17. Allen, *Practical forestry*, 3 (see chap. 2, n. 14).

18. Robert W. Steele, "Thinning nine-year-old Douglas fir by spacing and dominance methods," *Northwest Science* 29, no. 2 (1955): 84–89.

19. Thornton T. Munger, "Some forest problems of the Northwest," *Northwest Science* 2, no. 2 (1928): 42.

20. C. P. Willis and J. V. Hofmann, "A study of Douglas fir seed," *Proceedings of the Society of American Foresters* 10, no. 2 (1915): 142.

21. J. V. Hofmann, "Adaptation in Douglas fir," *Ecology* 2, no. 2 (1921): 131.

22. Jim Lichatowich, "Salmon without rivers: A history of the Pacific salmon crisis" (Washington, DC: Island Press, 1999), 128.

23. Roy R. Silen, John C. Weber, Donald L. Olson, and Ray J. Steinhoff, "Late-rotation trends in variation among provenances in four pioneer studies of ponderosa pine (*Pinus ponderosa Dougl. Ex Laws*)" (manuscript in preparation, on file, USDA Forest Service, Pacific Northwest Research Station, Corvallis, OR, 2003).

24. William G. Morris, notes to the FS 161 file on the Heredity Study, July 6, 1933 (on file, USDA Forest Service, Pacific Northwest Research Station, Forestry Sciences Lab, Corvallis, OR).

25. USDA Forest Service, Pacific Northwest Forest and Range Experiment Station, "The 1912 Heredity Study," USDA Forest Service, Pacific Northwest Research Station Annual Report (on file, Portland and Corvallis, OR, 1963), 4–7.

26. USDA Forest Service, "The principal laws relating to Forest Service activities," USDA Forest Service Agriculture Handbook 453 (Washington, DC: Government Printing Office, 1983), 5.

27. A. A. Griffin, "Influence of forests in melting of snow in the Cascade Range," *U.S. Monthly Weather Review* 46 (1918): 324–27.

28. W. E. Bullard, "Some references on watershed management," Research Note 63 (USDA Forest Service, Pacific Northwest Forest and Range Experiment Station, Portland, OR, 1950) 4.

29. Dennis R. Harr, "Effects of clearcutting on rain-on-snow runoff in western Oregon: A new look at old studies," *Water Resources Bulletin* 19, no. 3 (1986): 383–93.

30. David A. Clary, *Timber and the Forest Service* (Lawrence, KS: University of Kansas Press, 1986), 69.
31. J. V. Hofmann, "Natural regeneration of Douglas fir in the Pacific Northwest," USDA Department Bulletin 1200 (Washington, DC: Government Printing Office, 1924); Rodgers, *Bernhard Eduard Fernow*, 425 (see chap. 1, n. 5).
32. Allen, *Practical forestry*, 46.
33. Munger, *Growth and management* (see chap. 2, n. 15).
34. Hofmann, "Natural regeneration," 55.
35. Ibid.
36. J. V. Hofmann, "Natural reproduction from seed stored in the forest floor," *Journal of Agriculture Research* 11, no. 1 (1917): 5.
37. Rajala, *Clearcutting the Pacific rain forest*, 102 (see chap. 1, n. 12).

Chapter Four

1. Earle Clapp, *A national program of forest research* (Washington, DC: American Tree Association for the Society of American Foresters, 1926), 7.
2. Leo A. Isaac, "Douglas fir research in the Pacific Northwest, 1920–1956: An interview conducted by Amelia Fry" (Forest History Society, Berkeley, CA, 1967), 49.
3. Ibid., 52.
4. Robert W. Cowlin, "Federal forest research in the Pacific Northwest" (manuscript on file, USDA Forest Service, Pacific Northwest Research Station, Portland, OR, 1988) 24.
5. Robbins, *Lumberjacks and legislators*, 110 (see introduction, n. 18).
6. Isaac, "Douglas fir research," 50.
7. Ibid., 49.
8. Ibid.
9. Ibid., 50.
10. Ibid., 51.
11. William D. Miller, comp. "The Hofmann Trust: A history of the North Carolina Forestry Foundation" (Raleigh, NC: Hofmann Trust; North Carolina Forestry Foundation, North Carolina State University, 1970), 1–23.
12. Isaac, "Douglas fir research," 56–57.
13. Thornton T. Munger, "Objectives of the new federal forest experiment station," *Timberman* 26, no. 1 (November 1924): 54–55.
14. Thornton T. Munger, "The O.A.C. fernhopper's laboratory: The George W. Peavy Arboretum," *Oregon Agricultural College Annual Cruise* 7 (1926): 9.

15. Thornton T. Munger, "Ecological aspects of the transition from old forests to new," *Science* 72, no. 1866 (1930): 331.

16. Munger, *Growth and management*, 6 (see chap. 2, n. 15).

17. Wertz, "Record concerning the Wind River Forest Experiment Station," 13 (see chap. 3, n. 10).

18. Roy Silen, interview with the authors, 12 November 1999, on file, USDA Forest Service, Pacific Northwest Research Station, Corvallis, OR.

19. Hofmann, "Natural reproduction," 1–26 (see chap. 3, n. 36); Hofmann, "Natural regeneration" (see chap. 3, n. 31).

20. Leo A. Isaac, "Life of Douglas fir seed in the forest floor," *Journal of Forestry* 33, no. 1 (1935): 61–66.

21. Isaac, "Douglas fir research," 67.

22. Leo A. Isaac, "Seed flight in the Douglas fir region," *Journal of Forestry* 28, no. 4 (1930): 492–99.

23. R. E. McArdle and Leo A. Isaac, "Ecological aspects of natural regeneration of Douglas fir in the Pacific Northwest," *5th Pacific Science Conference Proceedings* 4 (1933): 4009–4015.

24. Thornton T. Munger, "Five years' growth on Douglas fir sample plots," *Proceedings, Society of American Foresters* 10 (1915): 423.

25. R. E. McArdle, "New yield tables for Douglas fir," *West Coast Lumberman* 50 (May 1, 1926): 100.

26. Ibid., 105.

27. Richard E. McArdle and Walter H. Meyer, "The yield of Douglas fir in the Pacific Northwest," USDA Forest Service Technical Bulletin 201 (1930; rev., Washington, DC: Government Printing Office, 1961).

28. Robert O. Curtis and D. D. Marshall, "A history of Bulletin 201" (talk, Western Mensurationist Meeting, Portland, OR, 13 June 2002).

29. McArdle and Meyer, "The yield of Douglas fir in the Pacific Northwest."

30. Leo A. Isaac, *10 years' growth of Douglas fir spacing-test plantations*, Research Note 23 (Portland, OR: USDA Forest Service, Pacific Northwest Forest and Range Experiment Station, 1937).

31. Munger, "Federal forest experiment station," 8.

32. Munger, "Forest research in the Pacific Northwest," 105 (see chap. 2, n. 1).

33. Ibid., 107.

34. Richard E. McArdle, "An outline of forest fire research in the Pacific Northwest," memorandum, 15 February 1927, on file, USDA Forest Service, Pacific Northwest Research Station, Portland, OR, 53.

35. William G. Morris, "Forest fires in western Oregon and western Washington," *Oregon Historical Quarterly* 35, no. 4 (1934): 313–39.

36. William Morris, *Diurnal changes in fuel moisture in the Douglas fir forest as affected by cover conditions,* Forest Research Note 23 (Portland, OR: USDA Forest Service, Pacific Northwest Forest and Range Experiment Station, 1937).

37. R. E. McArdle and Donald N. Mathews, *Fire research issue,* Forest Research Note 15 (Portland, OR: USDA Forest Service, Pacific Northwest Forest and Range Experiment Station, 1934); Donald N. Mathews, "Effects of shape, density, and methods of exposure of fuel moisture indicator sticks" (report on file, USDA Forest Service, Pacific Northwest Research Station, Portland, OR, 1940); William G. Morris, *Effect of ground surface and height of exposure upon fuel moisture indicator stick values,* Research Note 30 (Portland, OR: USDA Forest Service, Pacific Northwest Forest and Range Experiment Station, 1940).

38. A. G. Simson, "Research and the forest firefighter," *Four-L Lumber News* 8, no. 13 (May 1926), 55.

39. Donald N. Mathews, *Beware of rotten wood,* Research Note 27 (Portland, OR: USDA Forest Service, Pacific Northwest Forest and Range Experiment Station, 1939), 7.

40. Thornton T. Munger, "Snags," *Timberman* 28, no. 2 (1926): 37.

41. Thornton T. Munger, "Natural areas: Some historical notes" (report on file, USDA Forest Service, Pacific Northwest Research Station, Portland, OR, 5 March 1970).

42. I. W. Bailey and H. A. Spoehr, *The role of research in the development of forestry in North America* (New York: Macmillan, 1929), 16, 65, 103.

43. Harold K. Steen, *Forest Service research: Finding answers to conservation's questions* (Durham, NC: Forest History Society, 1998), 21; Samuel T. Dana and Sally K. Fairfax, *Forest and range policy: Its development in the United States* (New York: McGraw-Hill, 1980), 47; Robbins, *American forestry,* 48 (see chap. 1, n. 13).

44. Leo A. Isaac, *The Wind River Experimental Forest* (report on file, USDA Forest Service, Forestry Sciences Laboratory, Corvallis, OR, 1932), 12.

Chapter Five

1. Franklin and Eleanor Roosevelt Institute, "Franklin D. Roosevelt's Nomination Address, July 2, 1932, Chicago, IL," http://www.feri. org/ (accessed 23 December 2006).

2. Harold K. Steen, ed., *Forest and wildlife science in America: A history* (Durham, NC: Forest History Society, 1999), 7.

3. Clary, *Timber and the Forest Service,* 95–96 (see chap. 3, n. 30).

4. Edwin G. Hill, *In the shadow of the mountain* (Pullman, WA: Washington State University Press, 1990), 107–8.

5. Tolfree, "History of the Wind River Lumber Company, part two" (see introduction, n. 15).

6. Cheryl Mack and Rick McClure, "Significance evaluation of the Wind River administrative site historic district" (report on file, Heritage Program, Mt. Adams Ranger District, Gifford Pinchot National Forest, Trout Lake, WA, 1999), 15.

7. Hill, *In the shadow*, 108.

8. Felt, "Yacolt!" 31 (see introduction, n. 1).

9. Douglas C. Welch, *Pruning of selected crop trees in Douglas fir*, Forest Research Note 27 (Portland, OR: USDA Forest Service, Pacific Northwest Forest and Range Experiment Station, 1939); Leo A. Isaac, *Results of pruning to difference heights in young Douglas fir*, Forest Research Note 33 (Portland, OR: USDA Forest Service, Pacific Northwest Forest and Range Experiment Station, 1947).

10. Donald Worster, "O pioneers: Ecology on the frontier," in *Nature's economy: A history of ecological ideas* (Cambridge, UK: Cambridge University Press, 1985), 191–253.

11. Aldo Leopold, *A Sand County almanac and sketches here and there* (New York: Oxford University Press, 1987), 221.

12. Munger, "Some forest problems," 41 (see chap. 3, n. 19).

13. Schwantes, *Pacific Northwest*, 303 (see introduction, n. 9).

14. Thornton T. Munger, "Growth of Douglas fir trees of known seed source," USDA Technical Bulletin 537 (Washington, DC: Government Printing Office, 1936).

15. Hofmann, "Adaptation in Douglas fir" (see chap. 3, n. 21); William Morris, "Heredity tests of Douglas fir and their application to forest management," *Journal of Forestry* 32, no. 3 (1934): 351.

16. W. J. Allyn, "Climatological data 1911–1940, Wind River Station, Washington" (report on file, USDA Forest Service, Pacific Northwest Forest and Range Experiment Station, Portland, OR, 1941).

17. Thornton T. Munger and Ernest L. Kolbe, "The Wind River Arboretum from 1912 to 1932" (on file, USDA Forest Service, Pacific Northwest Forest and Range Experiment Station, Portland, OR, 1932); Thornton T. Munger and Ernest L. Kolbe, "The Wind River Arboretum from 1932 to 1937" (on file, USDA Forest Service, Pacific Northwest Forest and Range Experiment Station, Portland, OR, 1937).

18. Leo A. Isaac, *The effect of fire on Douglas-fir slash*, Research Note 3 (Portland, OR: Pacific Northwest Forest Experiment Station, 1929).

19. Leo A. Isaac and Howard G. Hopkins, "The forest soil of the Douglas fir region, and changes wrought upon it by logging and slash burning," *Ecology* 18, no. 2 (1937): 264–79.

20. Leo A. Isaac, "Factors effecting the establishment of Douglas fir seedlings," USDA Circular 486 (Washington, DC: Government Printing Office, 1938).
21. Ibid.
22. Isaac and Hopkins, "Forest soil."
23. Leo A. Isaac, "Vegetative succession following logging in the Douglas fir region with special reference to fire," *Journal of Forestry* 38, no. 9 (1940): 716–21.
24. Thornton T. Munger and Donald N. Mathews, "Slash disposal and forest management after clear cutting in the Douglas fir region," USDA Circular 586 (Washington, DC: Government Printing Office, 1941).
25. William G. Morris, *Influence of slash burning on regeneration, other plant cover, and fire hazard in the Douglas fir region (a progress report)*, Research Paper 29 (Portland, OR: USDA Forest Service, Pacific Northwest Forest and Range Experiment Station, 1958), 45.
26. Leo A. Isaac, "Fire, a tool, not a blanket rule in Douglas-fir ecology," in *Proceedings, second annual Tall Timbers fire ecology conference* (Tallahassee, FL: Tall Timbers Research Station, 1963), 16.
27. Leo A. Isaac, *Reproductive habits of Douglas fir* (Washington DC: Charles Lathrop Pack Foundation, 1943), 6.
28. Ibid., 105.
29. Allen, *Practical forestry*, 40–48 (see chap. 2, n. 14).
30. Burt P. Kirkland, "The need of a vigorous policy of encouraging cutting on the national forests of the Pacific coast," *Forestry Quarterly* 9, no. 3 (1911): 375–90.
31. Ibid., 376.
32. Cowlin, "Federal forest research," 63 (see chap. 4, n. 4).
33. Fernow, *Brief history of forestry*, 502 (see chap. 1, n. 1).
34. Clary, *Timber and the Forest Service*, 107.
35. Rajala, *Clearcutting the Pacific rain forest*, 127 (see chap. 1, n. 12).
36. Burt P. Kirkland and Axel J. F. Brandstrom, *Selective timber management in the Douglas fir region* (Washington, DC: USDA Forest Service, Division of Forest Economics, 1936), 35–47.
37. Rajala, *Clearcutting the Pacific rain forest*, 128.
38. Kirkland and Brandstrom, *Selective timber management*, 5.
39. Ibid., 4.
40. Ibid., 47.
41. Thornton T. Munger, "Practical application of silviculture to overmature stands now existing on the Pacific coast," *Pacific Science Congress Proceedings* 5 (1933): 4026.
42. Ibid.

43. Isaac, "Douglas fir seedlings."
44. Munger, "Forest research in the Pacific Northwest," 131 (see chap. 2, n. 1).
45. Isaac, "Douglas fir research," 89 (see chap. 4, n. 2).
46. Ibid., 91.
47. Kirkland and Brandstrom, *Selective timber management.*
48. Isaac, "Douglas fir research," 88.
49. Leo A. Isaac, *The place of partial cutting in old growth stands of the Douglas fir region*, Research Paper 16 (Portland, OR: USDA Forest Service, Pacific Northwest Forest and Range Experiment Station, 1956), 3.
50. Robert O. Curtis, "Selective cutting in Douglas fir," *Journal of Forestry* 96, no. 7 (1998): 44–45.
51. Robbins, *Landscapes of conflict*, 13 (see chap. 1, n. 11).

Chapter Six

1. Leo A. Isaac, "Biological aspects of forest conservation in Washington and Oregon," in *Conservation* (Corvallis, OR: Oregon State College, 1952), 12–15.
2. Vannevar Bush, *Science: The endless frontier* (Washington, DC: Government Printing Office, 1946), 5–6.
3. Robbins, *Landscapes of conflict*, 31 (see chap. 1, n. 11).
4. Hirt, *Conspiracy of optimism*, 131 (see introduction, n. 17).
5. Clary, *Timber and the Forest Service*, 125 (see chap. 3, n. 30).
6. Ibid., 158.
7. Philip Briegleb, *Applied forest management in the Douglas fir region*, Research Note 71 (Portland, OR: USDA Forest Service, Pacific Northwest Forest and Range Experiment Station, 1950).
8. Philip A. Briegleb and Kenneth H. Wright, "The development of forestry research on the forest lands of the state of Washington, 1940–1980," interview/discussion coordinated by Robert Torheim, 17 February 1989, transcribed and printed by Weyerhaeuser Company, University of Washington Libraries, Archives and Manuscripts Division, Seattle.
9. Robert W. Steele, comp., "Wind River climatological data, 1911–1950" (report on file, USDA Forest Service, Pacific Northwest Forest and Range Experiment Station, Portland, OR, 1954), 9.
10. R. W. Steele, *Cold weather damages promising species in the Wind River Arboretum*, Research Note 95 (Portland, OR: USDA Forest Service, Pacific Northwest Forest and Range Experiment Station, 1954).

11. Thornton T. Munger, *Recent growth records of Douglas fir stands*, Research Note 34 (Portland, OR: USDA Forest Service, Pacific Northwest Forest and Range Experiment Station, 1946), 9–10.

12. Thornton T. Munger, *Growth of ten regional races of ponderosa pine in six plantations*, Research Note 39 (Portland, OR: USDA Forest Service, Pacific Northwest Forest and Range Experiment Station, 1947).

13. Leo A. Isaac, *Better Douglas fir forests from better seeds* (Seattle: University of Washington Press, 1949).

14. Ibid., 6.

15. Leo A. Isaac, "Advantages of selecting seed trees with care" (talk, Western Forest Nurseryman's Meeting, Green Timbers British Columbia, 8 August 1952, on file USDA Forest Service, Pacific Northwest Research Station, Corvallis, OR).

16. Silen, interview with the authors (see chap. 4, n. 18).

17. A. E. Squillace and Roy R. Silen, "Racial variation in ponderosa pine," *Forest Science Monograph* 2 (1962).

18. Silen, Weber, Olson, and Steinhoff, "Late-rotation trends," 2 (see chap. 3, n. 23).

19. William Stein, interview with the authors, 13 March 2000, on file, USDA Forest Service, Pacific Northwest Research Station, Corvallis, OR.

20. Robert Tarrant, personal communication, 10 December 1999.

21. Ibid.

22. Tarrant, Robert F. "Attitudes toward red alder in the Douglas-fir region," in *Utilization and management of red alder*, General Technical Report 70, David C. Briggs, Dean S. DeBell, and William A. Atkinson, comp. (Ocean Shores, WA: USDA Forest Service, Pacific Northwest Forest and Range Experiment Station, 1978), 1–5.

23. Richard E. Miller and Marshall D. Murray, "The effects of red alder on growth of Douglas-fir," in Briggs, DeBell, and Atkinson, *Utilization and management of red alder*, 283–306 (see note 22 above); Robert F. Tarrant, "Stand development and soil fertility in a Douglas-fir–red alder plantation," *Forest Science 7*, no. 3 (1961): 238–46.

24. Robert F. Tarrant and Richard E. Miller, "Accumulation of organic matter and soil nitrogen beneath a plantation of red alder and Douglas-fir," *Soil Science Society of America Proceedings 27*, no. 2 (1963): 231–34.

25. Donald L. Reukema, *Growth response of 35-year-old site V Douglas-fir to nitrogen fertilizer*, Research Note 86 (Portland, OR: USDA Forest Service, Pacific Northwest Forest and Range Experiment Station, 1968).

26. Richard E. Miller and Donald Reukema, *Urea fertilizer increases growth of 20-year-old thinned Douglas-fir on a poor quality site*, Research Note 291 (Portland, OR: USDA Forest Service, Pacific Northwest Forest and Range Experiment Station, 1977).

27. R. F. Tarrant, H. J. Gratkowski, and W. E. Waters, *The future role of chemicals in forestry*, General Technical Report 6 (Portland, OR: USDA Forest Service, Pacific Northwest Forest and Range Experiment Station, 1973).

28. Donald A. Spencer, "Investigations in rodent control to advance reforestation by direct seeding: Progress report—spring 1951" (report on file, USDA Forest Service, Pacific Northwest Research Station, Forestry Sciences Lab, Corvallis, OR, 1951).

29. Ernest Wright and K. H. Wright, "Deterioration of beetle-killed Douglas-fir in Oregon and Washington: A summary of findings to date," Research Paper 10, Portland, OR: USDA Forest Service, Pacific Northwest Forest and Range Experiment Station, 1954).

30. T. W. Childs, "A case study of root and butt rot in second growth Douglas-fir" (report on file, USDA Bureau of Plant Industry, Division of Forest Pathology, USDA Forest Service, Pacific Northwest Research Station, Forestry Sciences Lab, Corvallis, OR, 1951).

31. Hirt, *Conspiracy of optimism*, 50.

32. Robbins, *Landscapes of conflict*, 22.

33. "Fabulous bear, famous service fight annual billion-dollar fire," *Newsweek*, June 2, 1952, 50–54.

34. Hirt, *Conspiracy of optimism*, 134–35.

Chapter Seven

1. Roy Silen, "Regeneration aspects of the 50-year-old Douglas fir heredity study," in *Proceedings of the 1964 Annual Meeting of Western Reforestation Coordinating Committee* (Portland, OR: Western Forestry and Conservation Association, 1965), 4.

2. Donald L. Reukema, "Final progress report on Wind River P.S.P.'s Martha Creek Flat and Lookout Mt. Road" (report on file, USDA Forest Service, Pacific Northwest Research Station, Portland, OR, 1987).

3. Donald L. Reukema, *Fifty-year development of Douglas-fir stands planted at various spacings*, Research Paper 253 (Portland, OR: USDA Forest Service, Pacific Northwest Forest and Range Experiment Station, 1979).

4. Richard E. Miller, Donald L. Reukema, and Harry W. Anderson, "Tree growth and soil relations at the 1925 Wind River spacing test in coast range Douglas-fir," Research Paper 558 (Portland, OR: USDA Forest Service, Pacific Northwest Research Station, 2004).

5. Donald L. Reukema, *Thirty-year progress report on the Planting Creek wide spacing trial (P18), Wind River Experimental Forest 1954–1984* (report on file, USDA Forest Service, Pacific Northwest Research Station, Portland, OR), 1988.

6. Constance A. Harrington and Donald L. Reukema, "Initial shock and long-term stand development following thinning in a Douglas-fir plantation," *Forest Science* 29, no. 1 (1983): 33–46.

7. Robert W. Steele, *Two commercial thinnings in century-old Douglas-fir*, Research Note 97 (Portland, OR: USDA Forest Service, Pacific Northwest Forest and Range Experiment Station, 1954).

8. Norman P. Worthington, *Lumber-grade recovery from 110-year-old Douglas-fir thinnings*, Research Note 121 (Portland, OR: USDA Forest Service, Pacific Northwest Forest and Range Experiment Station, 1955).

9. Richard L. Williamson, *Response to commercial thinning in a 110-year-old Douglas-fir stand*, Research Paper PNW-296 (Portland, OR: USDA Forest Service, Pacific Northwest Forest and Range Experiment Station, 1982).

10. Thomas W. Childs and Norman P. Worthington, *Bear damage to young Douglas-fir*, Research Note 113 (Portland, OR: USDA Forest Service, Pacific Northwest Forest and Range Experiment Station, 1955).

11. George Staebler, "Effects of controlled release on growth of individual Douglas-fir trees," *Journal of Forestry* 54, no. 2 (1956): 567–68.

12. Donald L. Reukema, "Progress report on the Planting Creek spot-thinning trial 1954–1983" (report on file, USDA Forest Service, Pacific Northwest Research Station, Olympia, WA, 1987).

13. Boris Zeide, "Thinning and growth: A full turnaround," *Journal of Forestry* 99, no. 1 (January 2001): 20–25.

14. Dean S. DeBell, Constance A. Harrington and John Shumway, "Thinning shock and response to fertilizer less than expected in young Douglas-fir stand at Wind River Experimental Forest," Research Paper PNW-RP-547 (Portland, OR: USDA Forest Service, Pacific Northwest Research Station, 2002).

15. Roy R. Silen and Donald L. Olson, *A pioneer exotic tree search for the Douglas-fir region*, General Technical Report 298 (Portland, OR: USDA Forest Service, Pacific Northwest Research Station, 1992).

16. Silen, "Regeneration aspects"; Squillace and Silen, "Racial variation in ponderosa pine" (see chap. 6, n. 17).

17. See note 1 above.

18. Frank H. Kaufert and William H. Cummings, *Forestry and related research in North America* (Washington, DC: Society of American Foresters, 1955).

19. Cowlin, "Federal forest research," 94 (see chap. 4, n. 4).

20. R. L. Williamson, *Growth and yield records from well-stocked stands of Douglas-fir*, Research Paper PNW-4 (Portland, OR: USDA Forest Service, Pacific Northwest Forest and Range Experiment Station, 1963), 1.

21. Donald L. Reukema, personal communication, 12 December 2000.

22. George S. Meagher to Harris, Furniss and TMR Project Leaders, memorandum, 20 May 1965, on file, USDA Forest Service, Pacific Northwest Research Station, Corvallis, OR.

23. Roy Silen to George S. Meagher, memorandum, 4 June 1965, on file, USDA Forest Service, Pacific Northwest Research Station, Corvallis, OR.

24. Silen, interview with the authors (see chap. 4, n. 18).

25. Ross Williams to director, PNW Forest & Range Exp. Station, memorandum, 29 October 1965, on file, USDA Forest Service, Pacific Northwest Research Station, Corvallis, OR.

26. Ross W. Williams, Forest Supervisor, to Regional Forester, R-6, and Director PNW Experiment Station, memorandum, 8 August 1967, on file, USDA Forest Service, Pacific Northwest Research Station, Corvallis, OR.

27. Ross W. Williams, Forest Supervisor, to director, Pacific Northwest Forest and Range Experiment Station, memorandum, 20 April 1966, on file, USDA Forest Service, Pacific Northwest Research Station, Corvallis, OR.

28. Roy Silen, Project Leader, to Dave Tackle, Assistant Director, memorandum, 12 July 1968, on file, USDA Forest Service, Pacific Northwest Research Station, Corvallis, OR.

29. Richard E. Miller, Project Leader Intensive Culture of Douglas-fir, to David Tackle, Assistant Director Timber Management Research, memorandum, 8 July 1968, on file, USDA Forest Service, Pacific Northwest Research Station, Corvallis, OR.

30. Thomas C. Adams, "Logging residues: Opportunities for greater utilization," (prepared for presentation at a meeting of the Portland Chapter, Society of American Foresters, Portland, OR, January 17, 1972, on file, USDA Forest Service, Pacific Northwest Research Station, Corvallis, OR).

31. Thomas Adams, *Logging costs for a trial of intensive residue removal*, Research Note 347 (Portland, OR: USDA Forest Service, Pacific Northwest Forest and Range Experiment Station, 1980); Thomas Adams, *Managing logging residue under the timber sale contract*, Research Note 348 (Portland, OR: USDA Forest Service, Pacific Northwest Forest and Range Experiment Station, 1980).

32. Constance A. Harrington and Dean S. DeBell, "The Trout Creek Hill research plantations, part I: The establishment phase" (report on file, USDA Forest Service, Pacific Northwest Research Station, Corvallis, OR, 1979).

33. Fernow, *Report upon the forestry investigations* (see chap. 1, n. 2).

34. Harrington and DeBell, "Trout Creek Hill research plantations."

35. Jack Rothacker, Research Project Leader, to R. W. Harris, Assistant Director, memorandum, 6 October 1969, on file, USDA Forest Service, Pacific Northwest Research Station, Corvallis, OR.

36. J. A. Kendall Snell and Timothy A. Max, "Estimating the weight of crown segments for old-growth Douglas-fir and western hemlock," Research Paper 329 (Portland, OR: USDA Forest Service, Pacific Northwest Forest and Range Experiment Station, 1985).

37. Jerry Williams, Historian, USDA Forest Service, Washington DC, personal communication.

Chapter Eight

1. Charles Connaughton, "What's ahead in forestry—a challenge to research," *Journal of Forestry* 69, no. 9 (1971): 556.

2. Samuel P. Hays, *Beauty, health, and permanence: Environmental politics in the United States, 1955–1985* (Cambridge, UK: Cambridge University Press, 1987), 13.

3. Clary, *Timber and the Forest Service*, 147–168 (see chap. 3, n. 30).

4. Jerry F. Franklin and Dean S. DeBell, "Effects of various harvesting methods on forest regeneration," in *Even-age Management Symposium*, Paper 848, ed. Richard K. Hermann and Denis P. Lavender (Corvallis, OR: Oregon State University School of Forestry, 1973), 29.

5. Dean DeBell, personal communication, 11–12 December 2000.

6. Jerry Franklin, personal communication, 5–6 March 2001; Robbins, *Landscapes of conflict*, 185 (see chap. 1, n. 11).

7. Clary, *Timber and the Forest Service*, 192–93.

8. See note 5 above.

9. Franklin, personal communication (see note 6 above).

10. Ibid.

11. Frank Benjamin Golley, *A history of the ecosystem concept in ecology* (New Haven, CT: Yale University Press, 1993), 118.

12. William R. Bentley, "Knowing ourselves: Changing definitions of the forestry profession," *Journal of Forestry* 93, no. 1 (1995): 12–15.

13. Leon Minckler, "Directions of forest research in America," *Journal of Forestry* 74, no. 4 (1976): 212, 214.

14. USDA Forest Service, "Timber cut and sold on national forests" (on file, Forest Service automated timber sales accounting system, USDA Forest Service regional office, Portland, OR).

15. Jerry Franklin, Chief Plant Ecologist, to Ed Clarke, Forest Residues Program, and Dean DeBell, Douglas-fir Silviculture, memorandum, 29 September 1977, on file, USDA Forest Service, Pacific Northwest Research Station, Corvallis, OR.

16. Robert E. Buckman, Deputy Chief of Research, to Directors, PNW, PSW, INT, and RM, memorandum, 29 July 1979, on file, USDA Forest Service, Pacific Northwest Research Station, Corvallis, OR.

17. Jerry F. Franklin, Chief Plant Ecologist, to R. M. Romancier, AD South, memorandum, 27 September 1979, on file, USDA Forest Service, Pacific Northwest Research Station, Corvallis, OR.

18. Robert F. Tarrant, Director, to E. H. Clarke, memorandum, 10 September 1979, on file, USDA Forest Service, Pacific Northwest Research Station, Corvallis, OR.

19. R. H. Waring and J. F. Franklin, "Evergreen coniferous forests of the Pacific Northwest," *Science* 204, no. 4425 (1979): 1380–86.

20. Jerry F. Franklin, personal communication, 3 October 2002.

21. Leopold, *Sand County almanac* (see chap. 5, n. 11).

22. Leonard F. Ruggerio, Keith B. Aubrey, Andrew B. Carey, and Mark H. Huff, eds., *Wildlife and vegetation of unmanaged Douglas-fir forests*, General Technical Report PNW-285 (Portland, OR: USDA Forest Service, Pacific Northwest Research Station, 1991).

23. Richard O. Woodfin, Dean S. DeBell, and Jerry F. Franklin, "Wind River Experimental Forest Management Plan" (report on file, USDA Forest Service, Pacific Northwest Research Station, Portland, OR, 1987), 147.

24. Ibid., 150.

25. Ibid., 152.

26. Jerry F. Franklin and Dean S. DeBell, "Thirty-six years of tree population change in an old-growth *Pseudotsuga-Tsuga* forest," *Canadian Journal of Forest Research* 18, no. 5 (1988): 633–39; Dean S. DeBell and Jerry F. Franklin, "Old-growth Douglas-fir and western hemlock: A 36-year record of growth and mortality," *Western Journal of Applied Forestry* 2, no. 4 (1987): 11–114.

Chapter Nine

1. Eric Forsman, Charles E. Meslow, and Howard M. Wright, "Distribution and biology of the spotted owl in Oregon," *Wildlife Monographs* 87 (April 1984): 1–64.

2. Eric Forsman, "Habitat utilization by spotted owls in the west-central Cascades of Oregon" (PhD diss., Oregon State University, 1980).

3. Ted Gup, "Owl vs Man," *Time*, June 25, 1990, http://www.time.com/time/magazine/article/0,9171,970447,00.html.

4. R. O. Curtis and A. B. Carey, "Timber supply in the Pacific Northwest: Managing for economic and ecological values in Douglas-fir forest," *Journal of Forestry* 94, no. 9 (1996): 6.

5. Robbins, *Landscapes of conflict*, 178–212 (see chap. 1, n. 11).

6. Hays, *Beauty, health, and permanence*, 255 (see chap. 8, n. 2).

7. K. Norman Johnson, Frederick Swanson, Margaret Herring, and Sarah Greene, eds., *Bioregional assessments: Science at the crossroads of management and policy* (Washington, DC: Island Press, 1999), 16.

8. See note 5 above.

9. David J. Brooks and Gordon E. Grant, "New approaches to forest management: Background, science issues, and research agenda; Parts one and two," *Journal of Forestry* 90, no. 2 (1992): 21–28.

10. Ruggerio, Aubrey, Carey, and Huff, *Wildlife and vegetation*, 2 (see chap. 8, n. 22).

11. Jerry Franklin, "Toward a new forestry," *American Forests* 95, nos. 11–12 (November/December 1989): 37–40.

12. Dean S. DeBell and Robert O. Curtis, "Silviculture and New Forestry in the Pacific Northwest," *Journal of Forestry* 91, no. 12 (1993): 26–30.

13. Ibid.

14. Johnson, Swanson, Herring, and Greene, *Bioregional assessments*, 12.

15. Jerry F. Franklin, "Scientists in wonderland," *Bioscience* 45, nos. 5–6 (1995): S74–S78.

16. K. N. Johnson, J. F. Franklin, J. W. Thomas, and J. Gordon, *Alternatives for management of late-successional forests of the Pacific Northwest*, report to the Agriculture Committee and the Merchant Marine Committee of the U.S. House of Representatives (Corvallis, OR: Oregon State University College of Forestry, 1991).

17. Johnson, Swanson, Herring, and Greene, "Learning from the past and moving to the future," in *Bioregional assessments*, 11–27.

18. Robbins, *Landscapes of conflict*, 210.

19. USDA and USDI, *Final supplemental environmental impact statement on management of habitat for late successional species and old growth forest related species within the range of the northern spotted owl, volumes 1 and 2* (Portland, OR: USDA Forest Service Regional Office, 1994).

20. USDA and USDI, *Record of decision for amendments to Forest Service and Bureau of Land Management planning documents within the range of the northern spotted owl* (Portland, OR: USDA Forest Service Regional Office, 1994), 6.

21. Gary Benson, Research and Monitoring Group, to Tom Mills, Station Director, memorandum, 18 July 1997, on file, USDA Forest Service, Pacific Northwest Research Station, Portland, OR.

22. Badege Bishaw, Dean S. DeBell, and Constance A. Harrington, "Patterns of survival, damage, and growth for western white pine in a 16-year-old spacing trial in western Washington," *Western Journal of Applied Forestry* 18, no. 1 (2003): 335–43.

23. Constance Harrington, personal communication, 2004.

24. Karl R. Buermeyer and Constance A. Harrington, "Fate of overstory trees and patterns of regeneration 12 years after clearcutting with reserve trees in southwest Washington," *Western Journal of Applied Forestry* 17, no. 2 (2002): 78–85.

25. See note 23 above.

26. Information on file, Mt. Adams Ranger District, Gifford Pinchot National Forest, Trout Lake, WA.

Chapter Ten

1. Brown alumni online, "The evolution of Nalini Nadkarni," *Brown University*, http://www.brown.edu/administration/brown_alumni_ magazine/95/7-95/features/nadkarni.html (accessed 9 December 2006).

2. DeBell, personal communication (see chap. 8, n. 5).

3. William G. Denison, "Life in tall trees," *Scientific American* 228, no. 6 (1973): 74–80.

4. Edward O. Wilson, "Rain forest canopy: The high frontier," *National Geographic* 180, no. 6 (1991): 78–107.

5. Franklin, personal communication (see chap. 8, n. 6).

6. David Shaw, personal communication, 2005.

7. Yves Basset, Vibeke Horlyck, and S. Joseph Wright, *Studying forest canopies from above: The international canopy crane network* (Panama City: Smithsonian Tropical Research Institute, 2003), 63–155.

8. S. C. Thomas and W. E. Winner, "Leaf area index of an old-growth Douglas-fir forest: An estimate based on direct structural measurements in the canopy," *Canadian Journal of Forest Research* 30, no. 12 (2000): 1922–30.

9. H. Ishii and E. D. Ford, "The role of epicormic shoot production in maintaining foliage in old *Pseudotsuga menziesii* (Douglas-fir) trees," *Canadian Journal of Botany* 79 (2001): 251–64.

10. William E. Winner, Sean C. Thomas, Joseph A. Berry, Barbara J. Bond, Clifton E. Cooper, Thomas M. Hinckley, James R. Ehleringer, et al., "Canopy carbon gain and water use: Analysis of old-growth conifers in the Pacific Northwest," *Ecosystems* 7, no. 5 (2004): 482–97.

11. Barbara Bond, personal communication.
12. Woods Hole Research Center, "The warming of the earth," http:// www.whrc.org/resources/online_publications/warming_earth/index. htm (accessed 19 December 2006).
13. Kyaw Tha Paw U, Matthias Falk, Thomas H. Suchanek, Susan L. Ustin, Jiquan Chen, Young-San Park, William E. Winner, et al., "Carbon dioxide exchange between an old-growth forest and the atmosphere," *Ecosystems* 7, no. 5 (2004): 513–24.
14. Mark E. Harmon, Ken Bible, Michael G. Ryan, David C. Shaw, H. Chen, Jeffrey Klopatek, and Xia Li, "Production, respiration, and overall carbon balance in an old-growth *Pseudotsuga-Tsuga* forest ecosystem," *Ecosystems* 7, no. 5 (2004): 498–512.
15. J. C. Domec, J. W. Warren, F. C. Meinzer, J. R. Brooks, and R. Coulombe, "Native root xylem embolism and stomatal closure in stands of Douglas-fir and ponderosa pine: Mitigation by hydraulic redistribution," *Oecologia* 141, no. 1 (2004): 7–16.
16. D. C. Shaw and S. B. Weiss, "Canopy light and the distribution of hemlock dwarf mistletoe (*Arceythobium tsugense (Rosendahl) G. N. Jones ssp. Tsugense*) aerial shoots in an old-growth Douglas-fir/western hemlock forest," *Northwest Science* 74, no. 4 (2000): 306–315.
17. See note 6 above.
18. B. McCune, R. Rosentreter, J. M. Pnzetti, and D. C. Shaw, "Epiphyte habitats in an old conifer forest in western Washington, U.S.A," *Bryologist* 103, no. 3 (2000): 417–27.
19. D. C. Shaw, E. A. Freeman, and C. Flick, "The vertical occurrences of small birds in an old-growth Douglas-fir–western hemlock forest stand," *Northwest Science* 76, no. 4 (2002): 322–34.
20. T. D. Scholwalter and L. M. Ganio, "Vertical and seasonal variation in canopy arthropod communities in an old-growth conifer forest in southwestern WA, USA," *Bulletin of Entomological Research* 88, no. 6 (1998): 633–40.
21. Zeide, "Thinning and growth," 22 (see chap. 7, n. 13).
22. USDA Forest Service, Pacific Northwest Research Station, "100,000 trees can't be wrong: Permanent study plots and the value of time," *Science Findings*, no. 64 (June 2004).
23. Andrew N. Gray and Thomas A. Spies, "Gap size, within-gap position and canopy structure effects on conifer seedling establishment," *Journal of Ecology* 84 (1996): 635–45; Robert Van Pelt and Jerry F. Franklin, "Response of understory trees to experimental gaps in old-growth Douglas-fir forests," *Ecological Applications* 9, no. 2 (2000): 504–512.
24. Isaac, *Reproductive habits of Douglas fir* (see chap. 5, n. 27).

25. J. F. Franklin, L. A. Norris, D. R. Berg, and G. Smith, "The history of DEMO: An experiment in regeneration harvest of northwest forest ecosystems," in "Retention harvests in northwestern forest ecosystems: The Demonstration of Ecosystem Management Options (DEMO) study," special issue, *Northwest Science* 73 (1999): 3–11.
26. USDA Forest Service, Pacific Northwest Research Station, "Developing silvicultural regimes: The eyes have it," *Science Findings*, no. 21 (January 2000).
27. USDA Forest Service, Pacific Northwest Research Station, "If you take a stand, how can you manage the complex art of raising a forest?" *Science Findings*, no. 27 (September 2000).
28. Silen and Olson, *Pioneer exotic tree search* (see chap. 7, n. 15).
29. Ibid., 1

Afterword

1. Cotta, "Cotta's preface" (see chap. 1, n. 3).
2. Ibid.
3. Tharandt Forest, http://www.forstbotanik.com (accessed 19 December 2006).
4. See note 1 above.

Index